U0363353

Research on Wild Green Peafowls in China

陈明勇　杨晓君　著

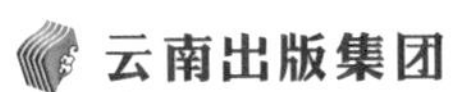
云南出版集团
Yunnan Publishing Group
云南科技出版社
Yunnan Science and Technology Press
·昆 明·
· Kunming ·

图书在版编目（CIP）数据

中国野生绿孔雀研究：汉、英 / 陈明勇，杨晓君著．
--昆明：云南科技出版社，2021.12
ISBN 978-7-5587-3992-7

Ⅰ．①中… Ⅱ．①陈… ②杨… Ⅲ．①野生动物—孔雀属—研究—中国—汉、英 Ⅳ．① Q959.7

中国版本图书馆 CIP 数据核字 (2021) 第 278696 号

• 国家出版基金资助项目
• 云南省科技厅科技人才与平台计划项目（202105AM070001）
亚洲象云南省野外科学观测研究站、
云南亚洲象教育部野外科学观测研究
• 站、生态环境部生物多样性调查、观测和评估项目
（2019—2023年）思茅区、江城县等
• 2区县生物多样性调查与评估（鸟类）项目
新平县野生绿孔雀现状调查项目

中国野生绿孔雀研究（汉、英）

ZHONGGUO YESHENG LÜKONGQUE YANJIU (HAN、YING)

陈明勇　杨晓君　著

出 版 人：温　翔
策　　划：温　翔　高　亢　刘　康　李　非　胡凤丽
责任编辑：王首斌　唐　慧　张羽佳　杨　雪　曹爱平　戴　熙　杨　楠　吴　琼
整体设计：长策文化
责任校对：秦永红　张舒园
责任印制：蒋丽芬
翻　　译：王海峰　于　萍　闫树蕙　郑乃齐　陈子倩 @ 语言桥
翻译校对：Brendan Hayes　张　义 @ 语言桥

书　　号：ISBN 978-7-5587-3992-7
印　　刷：昆明亮彩印务有限公司
开　　本：889mm × 1194mm　1/16
印　　张：26
字　　数：560 千字
版　　次：2021 年 12 月第 1 版
印　　次：2021 年 12 月第 1 次印刷
印　　数：1 ~ 3000 册
定　　价：100.00 元
出版发行：云南出版集团　云南科技出版社
地　　址：昆明市环城西路 609 号
电　　话：0871-64192760

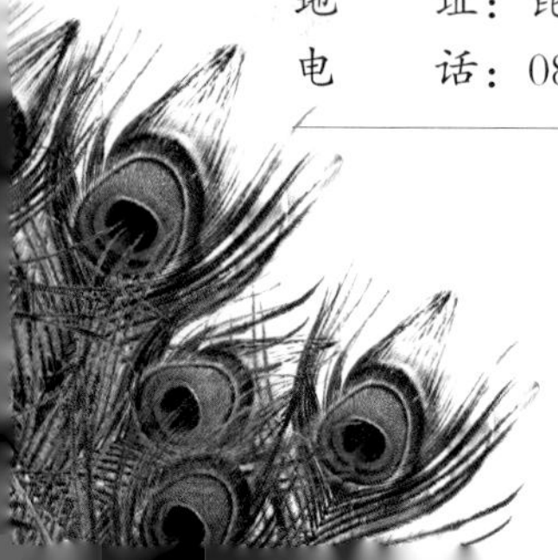

编委会名单

目录

CONTENTS

1

2

3

4

5

6

前言
FOREWORD

绿孔雀（*Pavo muticus* L.）是热带和亚热带低海拔地区分布的大型雉类。目前野生绿孔雀在世界上分布于缅甸、老挝、越南、泰国、柬埔寨、孟加拉国、泰国、马来西亚和印度尼西亚的爪哇岛等地，被世界自然保护联盟（IUCN）列为濒危（EN）物种，在《濒危野生动植物种国际贸易公约》（CITES）中被列入附录Ⅱ。

在我国历史上，野生绿孔雀曾广泛分布于长江流域及以南地区，包括湖北、湖南、四川、广东、广西、云南等省（自治区）。但由于人类捕杀、栖息地退化、人为干扰加剧等原因，野生绿孔雀的栖息地不断丧失，分布范围不断缩小，数量也在不断减少，甚至濒临灭绝。由于种群数量少、分布区域狭窄，绿孔雀在1989年就被列为首批国家Ⅰ级重点保护野生动物之一。然而，如此美丽、大型的鸟中“巨人”，它们的生存状况并不乐观，它们正在面临着生与死的炙烤！目前，野生绿孔雀在中国仅云南省北纬25°以南的少数地区还有遗存，总数量约550～600只，种群小而分散，仍然处于极度濒危状态；如果缺乏科学有效的保护，不久的将来极有可能在中国大地上灭绝。中国野生绿孔雀已然成为濒危物种，对它的保护、研究就显得尤为重要。保护生态环境，还绿孔雀一个生存家园已经刻不容缓！为了更好地保护野生绿孔雀，云南省还将其列为极危物种和极小种群物种，对其进行重点

拯救和保护。

绿孔雀体形硕大，体态矫健，羽毛华丽，姿态优美而高雅，在人类文明发展过程中被赋予深厚的文化内涵，是中国传统文化的重要组成部分，更是云南著名的文化、旅游名片。诞生于云南本土的以孔雀为原型的著名舞蹈“孔雀舞”“雀之灵”等更是享誉海内外，是中国文化的重要组成部分。在野生绿孔雀分布区，各族人民都十分喜爱绿孔雀，也产生了丰富的孔雀文化。尤其在云南的西双版纳、德宏等傣族聚居区，绿孔雀是傣族人民心目中吉祥、幸福的象征，有着许多优美的故事、神话与传说。孔雀文化与傣族文化之间的关系更是渊源深厚、光辉灿烂。因此，绿孔雀独有的文化价值同样需要被保护，保护绿孔雀的同时，也是在保护中国灿烂的孔雀文化。

文献资料显示，目前国内还没有一本专门介绍我国野生绿孔雀的书籍，对它们的研究文献也寥寥无几。如此薄弱的研究，与绿孔雀本身的光彩夺目形成了极大的反差！我们对中国野生绿孔雀的认识和了解是十分有限的，就比如说，野生绿孔雀的巢是什么样的？建在什么地方？雏鸟是怎么生活的？绿孔雀的食物都有什么？除了人以外，它们面临的天敌还有哪些？蓝孔雀与绿孔雀杂交会造成野生绿孔雀基因污染吗？怎样才能做好野生绿孔雀的科学保护与管理？……这些基本的科学问题都还需要深入研究。在此基础上，积极对它们开展有效的保护、管理就显得非常重要了。

保护中国野生绿孔雀及其栖息地，需要国际、国家层面予以高度重视，更需要云南省、各相关州（市）、分布区各县政府及各相关部门的重视与努力，以及社会各界的共同关注和通力合作。政府及各部门的宏观决策需要系统、科学的数据支撑，然而对我国野生绿孔雀的研究却十分薄弱，更缺乏完整而系统的科普读物，使政府及各部门领导在决策时面临极大的困难，社会各界也无从认识和了解绿孔雀这一具有传奇色彩的珍稀鸟类。缺乏相关资料和书籍，公众对野生绿孔雀世界的真实理解就缺少必要的资源，在文化传承方面，我们也将失去“根”与“土

壤”，也就无法生根发芽，开花结果。

云南大学在生物学、生态学的学科发展中有十分悠久的历史，长期以来积淀了大量的研究成果。在鸟类学的研究中，几代科研工作者付出了辛勤的汗水，取得了大量的研究成果，并培养了大批科学工作者，其中不乏像王紫江、江望高等著名鸟类学家，也涌现出了许多颇具实力的科研新秀，为云南大学在鸟类学研究工作中奠定了良好的基础。2016年初开始，受林业和草原管理部门、自然保护区管理部门的邀请和委托，云南大学先后承担了一些云南省野生绿孔雀调查研究工作。经过多年的努力，云南大学采用红外相机拍摄到大量珍贵的野生绿孔雀照片和视频，对野生绿孔雀的生境、行为、分布的研究有了长足的进展，获得了丰硕的成果，开启了云南大学绿孔雀研究方面的新篇章。

云南大学动物多样性保护研究团队通过对本团队获得的大量第一手资料、数据进行系统整理，结合国内外研究成果和历史资料，组建云南大学绿孔雀研究团队，联合中国科学院昆明动物研究所杨晓君研究员共同编写了《中国野生绿孔雀研究》一书。本书介绍了中国野生绿孔雀形态特征、分布与现状、生存环境、食性及生境选择，以及它们采食、繁殖行为等生物学知识，结合收集到的我国各地少数民族文化中关于孔雀的民间传说、神话故事等颇具文化价值的资料，配以大量的彩色照片，图文并茂地呈现中国野生绿孔雀的生存及保护现状。本书的野外数据采集过程中，普永云（云南哀牢山国家级自然保护区新平管护局原局长）、刘德军（云南大学绿孔雀研究团队）、张宏雨（云南哀牢山国家级自然保护区新平管护局原保护科科长/科研所所长）、王方（云南大学生态与环境学院硕士研究生，已毕业）参与到编写中，野外数据的整理和筛选主要由王方和梁良完成。

本书将填补中国尚无野生绿孔雀专著的空白，弥补绿孔雀研究数据的不足，为今后绿孔雀的保护、科研、科普及教学提供基础资料，是一本对绿孔雀及其栖息地保护有价值的参考书。

Foreword

The green peafowl is a large pheasant distributed in tropical and subtropical low-altitude regions. The wild green peafowls are distributed in the countries such as Myanmar, Laos, Viet Nam, Thailand, Cambodia, Bangladesh, Thailand as well as Malaysia and the Java Island of Indonesia. It is recognized by International Union for Conservation of Nature (IUCN) as a global endangered (EN) species, and is listed in Appendix II in the Convention on International Trade in Endangered (EN) Species of Wild Fauna and Flora (CITES).

In the history of China, wild green peafowls were once widely distributed in the Yangtze River Basin and the south areas of it, including Hubei, Hunan, Sichuan, Guangdong, Guangxi, Yunnan, etc. However, due to human hunting, habitat degradation, and increased human disturbance, the habitats of wild green peafowl have been continuously lost, their distribution areas numbers have been declining. The wild green peafowls are even on the verge of extinction, and were listed first in 1989 as one of the national Class I key protected wild animals. However, the living conditions of such beautiful and "giant" birds are by no means promising. They are struggling between life and death. At present, there are still a small number of wild green peafowls in only a few south areas towards the latitude 25°N in Yunnan Province in China. It is estimated that the total number is 555~600. These small amounts of wild peafowls are sporadically scattered, and still critically endangered. Without scientific and effective protection, they may become extinct in China in the near future. The wild green peafowl in China has become an endangered species, and its protection and research are particularly important. Protecting the ecological environment and restoring the habitats of green peafowls are pressing tasks. In order to protect the wild green peafowl better, Yunnan Province has also listed it as a critically endangered species with a tiny population, and made great efforts to save and protect it.

The green peafowl, with a large size, vigorous posture, gorgeous feathers, and elegant postures, is endowed a deep cultural connotation in the development of human civilization and has become an important symbol in traditional Chinese culture. It is also a famous cultural and "tourist business card" in Yunnan. The famous dancing shows of "Dance of Peafowl" and "Spirit of Peafowl", which take inspiration from the peafowls in Yunnan, are well-known at home and abroad. In areas where wild green peafowls are distributed, people of all ethnic groups are very fond of them, and a large amount of green peafowl culture has also emerged. In the Dai ethnic minority areas such as Xishuangbanna and Dehong in Yunnan in particular, the green peafowl is a symbol of auspiciousness and happiness in the Dai people's heart. In addition, there are many beautiful stories, myths and legends about the peafowl. The peafowl culture and the Dai culture are closed bounded and complement with each other. Therefore, the unique cultural value of the green peafowl also needs to be protected.

Protecting the green peacock is also protecting China's splendid traditional national culture. Literature data shows that there is no book dedicated to introducing wild green peafowls in China, and there are very few research papers on them. Such weak studies are in great contrast to the brilliance of the green peafowls itself! Our knowledge and understanding of them, therefore, is very limited. What is the nest of wild green peafowls? Where is it built? How do young peafowls live? What is the food of the green peafowl? In addition to people, what other natural enemies do they have? Will the blue peafowl hybridizing with the green peafowl result in genetic pollution of the wild green peafowl? How can we do scientific protection and management of wild green peafowls? ...These basic scientific issues still need to be studied in depth. And on the basis of these questions and surveys, effective protection, management, and even utilization should be actively carried out.

Protecting the wild green peafowls and their habitats in China requires great attention at both the international level and the national level, the attention and efforts of Yunnan Province, all related cities, counties, and related departments, as well as the shared attention and efforts of all sectors of the society. The macro decision-making of the government and various departments requires systematic and scientific data support. However, there are limited researches on wild green peafowls in China, as well as complete and systematic popular science books. The government and various department leaders are difficult in making decisions, and people from all walks of life have no way of knowing and understanding the green peafowl, the rare and legendary bird. Without relevant materials and books that are widely available to the public, the public have no reliable resources to truly understand the world of wild green peafowls. In terms of cultural heritage, we also lose our "root" and "soil", and achievements in cultural heritage cannot be made.

Yunnan University has a very long history in the development of biology and ecology, and has made a large number of research results for a long time. In the research of ornithology, a lot of research achievements have been made by several generations of scientific researchers. It has cultivated a large number of scientific professionals, including famous ornithologists such as Wang Zijiang and Jiang Wanggao having trained many young talents in scientific research and laying a solid foundation for Yunnan University in ornithology research. Since the beginning of 2016, Yunnan University has undertaken some surveys and researches on wild green peafowls in Yunnan Province at the invitation and commission of the forestry and grassland management department and nature reserve management department. In the following four years, a large number of precious photos and videos have been taken by infrared cameras, and great progress in the study of the habitat, behavior and distribution of wild green peafowls has been made, opening a new chapter in the research of green peafowls at Yunnan University.

By systematically sorting out the large amount of first-hand materials and data obtained by the Animal Diversity Conservation Research Team of Yunnan University, and referring to domestic and foreign research results and historical data, the Green Peafowl Research Team of Yunnan University and the researcher Yang Xiaojun with Kunming Institute of Zoology, CAS compiled the book *Research on Wild Green Peafowls in China*, which has introduced the Chinese wild green peafowl's appearance characteristics, distribution and present situation, living environment, food habits and habitat selection, as well as biological knowledge of their feeding and reproductive behavior. Besides, a collection of folklore, mythical stories and other cultural materials about peafowls in ethnic minority cultures in China, and a large number of color photographs are added into the book, presenting the survival and protection status of China's wild green peafowls with both texts and photos in a direct and true–to–life manner. The field data collection of this book is mainly conducted by Pu Yongyun (former director of Xinping Management and Protection Bureau of Yunnan Ailao Mountain National Nature Reserve), Liu Dejun (member of the Green Peafowl Research Team of Yunnan University), Zhang Hongyu (former chief of the Conservation Section of Xinping Management and Protection Bureau of Yunnan Ailao Mountain National Nature Reserve, and director of the Scientific Research Institute), and Wang Fang (postgraduate of the Ecology and Environment College of Yunnan University, graduated). And the field data is mainly sorted and selected by Wang Fang and Liang Liang.

This book will fill absence of wild green peafowl monographs in China and make up for the lack of research data on peafowls, thus providing basic materials for future protection, scientific research, and science popularization about peafowls. It will be a valuable reference book for leaders of governments and departments at all levels, wildlife and nature conservation professionals, bird and green peafowl lovers, and the public.

1 中国野生绿孔雀的分布历史与现状

1.1　历史时期中国绿孔雀分布

1.1.1　关于河南省孔雀分布的讨论

据河南省淅川县下王岗遗址第九文化层（仰韶早期）中发现孔雀属（*Pavo* sp.）的遗骨（跗跖骨）……这是喜暖动物所占比例最多的时代，说明仰韶文化期是下王岗遗址最温暖时代的代表（贾兰坡和张振标，1977）。作者在文章中未说明这次发掘的孔雀遗骨是当地野生的孔雀还是家养的孔雀，也未说明是绿孔雀还是蓝孔雀的遗骨。我国历史地理学家文焕然和何业恒先生据此认为“据河南省淅川县下王岗遗址第九文化层中，发现孔雀属的遗骨，说明距今五六千年前，秦岭东南端天然森林与开阔地灌木的接触地带有野生绿孔雀分布”（文焕然和何业恒，1980，1981，2006；何业恒，1994）。由于河南是迄今为止查到文献记载的我国野生绿孔雀最北的分布区，为慎重起见，我们对《河南淅川县下王岗遗址中的动物群》一文认真阅读和分析后，发现文章中报道的仰韶早期这一文化层还发现的动物骨骼还有鳖属（*Trionyx* sp.）、猕猴（*Macaca mulatta*）、狗（*Canis familiaris*）、黑熊（*Selenarctos thibetanus*）、虎（*Panthena tigris*）、苏门犀（*Didermocerus sumatrensis*）、亚洲象（*Elephas maximus*）、野猪（*Sus scrofa*）、家猪（*Sus scrofa domesticus*）、麝（*Moschus moschiferus*）、麂（*Muntiacus* sp.）、斑鹿（*Cervus nippon*）等。作者在文末强调：“从下王岗发现的仰韶文化期的家畜来看，例如猪和狗已经和它们的祖先型——野猪和狼有了很大的区别，证明绝不是最初的饲养种，在仰韶文化期以前必然已经经过了长期的饲养阶段。”基于这些资料，我们认为，不能排除这次考古发现的孔雀属动物的遗骨（跗跖骨）是家养孔雀的可能性，将它作为野生孔雀的分布点，证据是不够充分的，值得商榷和进一步深入研究。

1.1.2 关于湖北、湖南孔雀分布

文焕然和何业恒（1981，2006）根据战国时《楚辞·大招》载“孔雀盈园”。汉王逸注“言园中之禽，则有孔雀群聚，盈满其中”，认为这里指的是饲养的孔雀。但却得出结论：“却反映距今2000多年前的楚国，约相当今湖北、湖南等地，可能有野生孔雀的分布。”我们认真分析后认为，有饲养的孔雀，不代表就有野生孔雀的分布，这一结论也是值得商榷的，还需要找到野生孔雀分布的更多证据，才能确定湖北、湖南等地有野生孔雀的分布。后来何业恒（1994）提供了更多的证据，清乾隆十九年（1754）和同治五年（1886）《长阳县志·物产》中有“孔雀”，1921年《湖北通志·物产》：“孔雀，长阳出，见县志（指1754年和1886年的长阳县还有野生孔雀分布）。”《图出集成·方舆汇编·职方典·永州府物产考》羽之属有：“鸳鸯、鹧鸪、锦鸡、白鹇、孔雀、鹦鹉等。”清初的永州府辖今永州（市）、零陵、双牌、道县、祁阳、东安、宁远、江水、江华、新田等县（市），这里的孔雀与锦鸡、白鹇一样，都属野生，而不是人工饲养的。乾隆三十年（1765）《辰州府志·物产》记载府内（今沅陵、辰溪、泸溪、溆浦等县）主要野生禽类有“孔雀”“鹤”“鹧鸪”，并且在“孔雀”中编者自注：“五色俱备，其尾长如屏，光艳可爱。”这些情况，反映18世纪中叶以前湖北、湖南地区还有野生孔雀分布，则古代这一带有野生孔雀存在，可以想见（何业恒，1994）。

1.1.3 关于浙江省孔雀分布的讨论

在18世纪前，浙江西南也有野生孔雀分布。康熙三十一年（1692）《金华府义乌县志·物产》中有“孔雀”，雍正五年（1727）《志》同，可以为证（何业恒，1994）。

1.1.4 关于新疆维吾尔自治区孔雀分布的讨论

文焕然和何业恒（1981，2006）在对历史文献的研究中发现，在我国西北的新疆塔里木盆地也有孔雀的记载。《太平御览》卷924载，“（三国）魏文帝与群臣诏曰：前于阗（今新疆和田）王所上孔雀尾万枚……”；《北史》卷97《西域·传龟兹国》北魏时龟兹（今新疆库车）“土多孔雀，群飞于谷间，人取养而食之，孳乳如鸡鹜，其王家恒有千余只云”，但还待验证。

王子今（2013）认为，如果西域的“孔雀”是来自天竺等地，则龟兹在中外交通史上的地位又有了新的证明。其实，现在尚不能完全排除当时龟兹地方原生“孔雀”的可能，因为史书最有力的记载：“土多孔雀，群飞于谷间，人取养而食之，孳乳如鸡鹜。”尚不能轻易否定。尽管现今这一地区已经看不到“孔雀”的踪迹，但是有可能今后的考古工作可以证明上古时代龟兹“孔雀”的实际存在，如同淅川的发现那样。我们从更早的历史迹象中也可以发现支持这一意见的信息，例如《逸周书·王会》说成周之会，四方朝于内者，“方人以孔鸟”（晋孔晁注：“亦戎别名”），反映当时已有西方部族进献“孔鸟”的情形。

白娜（2015）以《二十四史》为主要来源，辅助其他史籍，梳理分析与孔雀相关的历史文献，认为孔雀在中国历史时期主要分布在今天云南、广东、广西、新疆地区，其次是四川、湖南、湖北、海南、西藏地区，在中原河南偶有畜养。

1.1.5 关于中国境内孔雀的来源讨论

王研博（2018）、王研博和郭风平（2018）均认为，从世界范围内看，中国境内的孔雀基本来自外域，归类出两个方向：一是源于东南亚，包括中南半岛等，经陆路或水路从中南半岛扩散至中国的云南、贵州、广西、广东地区，并经由商人交易抑或使团进贡的方式传入中原。秦汉之际，栖息于东南亚的绿孔雀即已在中国岭南地区生活，但数量较少，故而可作为珍奇异宝进贡皇家或王公贵族。据史书载，一是庄蹻似与尉陀同时，大约在秦汉之际由楚地入滇。滇，即今云南、贵州地区，与中南半岛接壤，且往来已久。而此地又多出“孔雀”，可理解为此地本就有绿孔雀之原生种，但也可认为由南方路传来。二是源于印度半岛及斯里兰卡，经古丝绸之路，由南亚经中亚或翻越葱岭过新疆乃至河西走廊进入中原。在中古时期孔雀似乎已大量存在于西域地区。不过，栖息于此的到底是印度蓝孔雀还是爪哇绿孔雀，似乎无法从史料中判断。然而，从绿孔雀的生存条件来看，不管是气候还是地貌因素，西域龟兹似乎都无法满足。作为生存适宜能力更强，且在传播路径上更为符合现实的蓝孔雀应该才是这一地区真正的“孔雀”。

我们查阅了大量动物学研究的文献，分析后认为：

①该研究认为“中国境内的孔雀基本来自外域，且滇西南、广西、广东地区的孔雀来源于东南亚、中南半岛”的说法是错误的，云南许多地区目前还存在着许多原生的野生绿孔雀就是明证。

②由于古籍中并没有关于孔雀是蓝孔雀还是绿孔雀的描述，该文通过绿孔雀的生境需要推理在西域/新疆地区的孔雀应为蓝孔雀，其来源于印度半岛及斯里兰卡，这一推断具有一定的合理性。

1.1.6　中国野生绿孔雀历史分布

文焕然等（1981，2006）和何业恒（1994）均认为，历史时期中国的孔雀主要分布在长江流域及其以南地区，大体可分为如下三个区：长江流域、岭南和滇西南。各分布区分述如下：

（1）长江流域：包括滇东北、四川盆地及长江中游一带

长江上游的四川盆地和滇东北，是古代多孔雀的地方。晋代左思《蜀都赋》指出四川盆地“孔翠群翔，犀象竞驰”。唐刘良注：“孔，孔雀；翠、翠鸟也。”表明3世纪末，这一带的野生孔雀是不少的。

《后汉书·南蛮西南夷列传·滇》：益州郡“河土平敞、多出鹦鹉、孔雀、有盐池田渔之饶，金银畜产之富”。晋常璩《华阳国志·南中志》：“晋宁郡，本益州也。……治滇池（今云南晋宁晋城）……郡土大平敞，原田多长松皋，有鹦鹉、孔雀、盐池、田渔之饶”。这些材料，说明距今1000多年前，在云南东北一带是多野生孔雀的（何业恒，1994）。这里需要说明的是，现今的晋宁为昆明市的一个区，位于昆明主城区以南，并不属于滇东北，而属于滇中地区。

（2）岭南

根据古籍记载，按照孔雀的地理分布，在岭南可分为6个地区：粤东地区（广东省潮阳县）、粤中地区（广东省从化县）、云开大山及其附近地区（广东省徐闻县、廉江县、海康县、吴川县、高州，广西壮族自治区的玉林县、博白县）、桂北地区（广西壮族自治

区蒙山县、宜山县、临桂县、兴安县、灵川县、永福县、阳朔县）、桂西南地区（广西壮族自治区西武鸣县、宜州县、郁江县、红水河、左江、右江、郁江及黔江流域）、桂东南地区（广西壮族自治区南部六万大山至十万大山南北一带，即合浦县、钦县、时罗都）（文焕然等，1981，2006）。从西汉到南宋，孔雀在岭南一带为常见的飞禽（何业恒，1994），广西壮族自治区西部山地还有不少的孔雀。据清代人的著作《粤西偶记》当时广西壮族自治区的南宁、来宾、桂平一带，江河等流域，分布的野生孔雀是不少的（文焕然和何业恒，1980）。《盐铁论》亦说："南越以孔雀饵门户"，故《昊录・地理志》称："郡内及朱崖皆有之"，可见海南岛也是有孔雀的（曾昭漩，1980）。

（3）滇西南

滇西南是我国野生孔雀分布历史悠久的地区，据《华阳国志・南中志》："永昌郡，古哀牢国"，"土地沃腴，物产丰富"，有"孔雀、犀、象"等珍禽异兽。哀牢为古国名，东汉永平十二年（公元69年）以其地置哀牢（今云南盈江县东），博南（今永平县）两县，属永昌郡。当时永昌郡所辖的范围甚广，其地东西3000余里，南北4800余里（1里=500米），大体相当于今大理白族自治州、保山市、临沧市、德宏傣族景颇族自治州及西双版纳傣族自治州等地。从汉代以来，滇西南一直以产孔雀著名。例如，在唐朝的时候，云南省德宏傣族景颇族自治州一带"孔雀巢人家树上"，一般人家房前屋后的树上，都有孔雀在上面做巢，当时这一带孔雀之多，就可以想见了（文焕然和何业恒，1980，2006）。从汉代到元代一直有野生孔雀栖息（文焕然和何业恒，1980，1981，2006）。据明代隆庆六年（1572）《云南通志》和天启五年（1652）《滇志》，当时云南境内孔雀的分布范围，曾到达文山、开远、昆明、富民、楚雄等地。在建水县、大理市、保山市、梁河县及元江县、镇沅县、景东县、凤庆县等地野生绿孔雀也广泛分布（何业恒，1994）。到了清代，在昆明市、楚雄市、玉溪市（元江哈尼族彝族傣族自治县）、普洱市（镇沅彝族哈尼族拉祜族自治县、景东彝族自治县、宁洱哈尼族彝族自治县）、大理白族自治州（大理市、永平县）、临沧市（凤庆县）、保山市（隆阳区、腾冲市）、红河哈尼族彝族自治州（建水县）、西双版纳傣族自治州（景洪市）、德宏傣族景颇族自治州（芒市、梁河县、陇川县）有野生绿孔雀的分布记载。

根据文焕然和何业恒绘制的《中国孔雀分布变迁示意图》（图1-1）涉及的分布地点包括广东省、广西壮族自治区和云南省的部分县（市、区），具体地点如下：

广东省：广州（市）、潮阳县（今汕头市潮阳区）、罗定县（今罗定市）、高州县（今高州市）东北、吴川县（今吴川市）西南、廉江县（今廉江市）北、海康县

（今雷州市）、徐闻县。

广西壮族自治区：南宁市、武鸣县、上林县、蒙山县南、玉林市（今玉林市玉州区）西北、容县、博白县、钦州（市）、灵山县、合浦县、防城各族自治县（今防城港市）、崇左市（今崇左市江州区）、宜山县（宜州市）。

云南省：晋宁区、元江哈尼族彝族傣族自治县、保山市（今保山市隆阳区）、普洱（今宁洱哈尼族彝族自治县）、景东彝族自治县、镇沅彝族哈尼族拉祜族自治县、凤庆县、梁河县、盈江县、陇川县、永平县、景洪县（今景洪市）。

除了上述历史时期提到的我国野生绿孔雀分布在滇西南地区外，云南省其他地区的孔雀分布也有一些描述，如白娜（2015）认为滇中、滇西、滇东、滇南皆有孔雀分布的记载。

▲ 图1-1　中国孔雀分布变迁示意图（引自文焕然等，2006，有修改）

本图境界画法不作划界依据

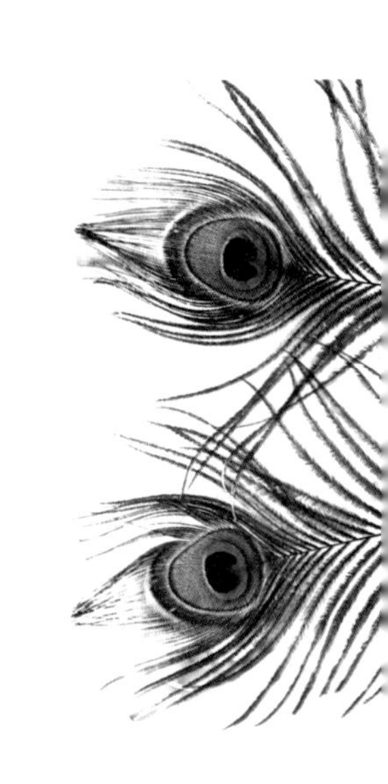

1.2　我国野生绿孔雀分布区历史性消退及地区性灭绝

1.2.1　中国野生绿孔雀分布区的历史性消退

蓝勇（2002）认为，历史时期中国野生孔雀分布从河南南部的北纬33.1°降至北纬25.4°，南移了7.7个纬度。文焕然和何业恒（1981，1995，2006）认为，各个历史时期中国孔雀的地理分布，从北向南、从东北到西南逐步缩小，目前云南西南部成为我国野生孔雀仅有的分布地区。我国地区开发时间，大体上是长江流域较早，珠江流域次之，最后为滇西南，这样就使得孔雀分布的地界和象、犀有相似之处，即逐渐南移。王研博和郭风平（2018）认为，从环境史的角度，对历史文献的整理分析，历史时期中国境内的孔雀从数量上呈现从北向南、终至西南的变迁特点。

1.2.2　岭南野生绿孔雀分布区的历史性消退

对于中国境内孔雀主要分布区的区域性变迁，国内学者已有相当的认识，如文焕然与何业恒在他们的研究论文和专著中提到：云开大山及其附近地区东北部孔雀的灭绝时期，似在18世纪30年代以后。桂北地区，宋时宜州（今宜山县）山中多孔雀，但以后就不见于记载了。18世纪以后，桂东南地区一带的野生孔雀已逐渐减少。总之，岭南各历史时期野生孔雀分布很广，灭绝的总趋势是北早南迟，东先西后，即广西、广东的北部和中部较南部为早，广东较广西为先；平原丘陵较早，山地较晚，广西东南部六万大山与十万大山之间的灵山县一带的野生孔雀，直到20世纪才趋于灭绝，岭南野生孔雀分布变迁的趋势也大致上反映了地区开发的总趋势（文焕然和何业恒，1981，1995，2006）。

1.2.3 珠江三角洲孔雀分布记录

华南（孔雀）分布区也很广泛，如《南越志》称："义宁县杜山多孔雀。"唐代《岭表录异》亦记："孔雀翠尾，自累其身。"可见当时广州地区是有孔雀的。《北户录》亦称："广之南、新、勤、春十州，多孔雀。"可见南海、新会地区在唐代和粤西山地一样有孔雀栖息。义宁亦即今开平，可知珠江三角洲附近山地在晋代以来，即为孔雀分布区，到唐代仍多。如《新会乡土志》（1908）说："唐贞元（785—850）时，杜枯为广州刺史，自言其治下新会桂山，有翡翠、孔雀、元猿。附郭之圭峰，犹有珍禽，其他山林，未尽启辟可知。"但经宋代后，在珠江三角洲区已少记录。南宋对孔雀大量捕杀记录见于《岭外代答》，该书卷九称："中州人得一，则储之金属，南方人乃腊而食之，物之贱于所产者如此。"这样，珠江三角洲地区宋后就少记录了。因此，珠江三角洲地区孔雀灭绝，比广西、粤西为早。据清代著作，广西、粤西还有记录，如《南越笔记》说："孔雀产高、廉、雷、罗定诸处。"《岭南杂记》说："孔雀产广西，而罗定山中间或有之。"可见广东产区已大为缩小。

1.2.4 滇西南野生绿孔雀的历史性消退

滇西南一带从汉代到元代一直有野生孔雀栖息。据文献记载，明清时期滇西南一带野生孔雀的分布，大体限于元江、镇沅、景东、凤庆、保山一线以南、以西地区，元江、镇沅、景东等县在清代都产孔雀，而现今却已灭绝，可见滇西南野生孔雀的分布范围，又有进一步的缩小（文焕然和何业恒，1981，1995，2006）。

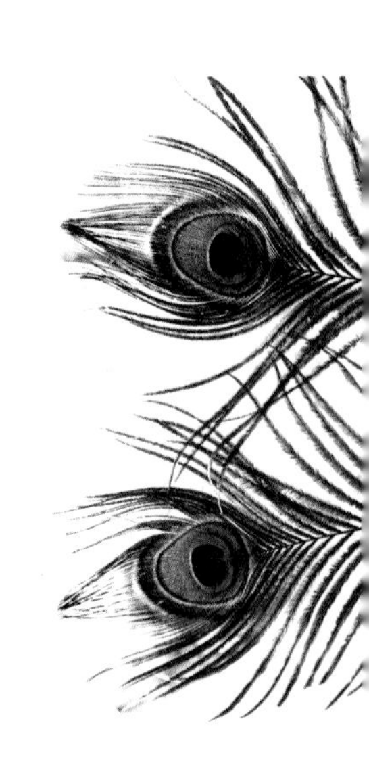

1.3　中国野生绿孔雀分布变迁的原因

文焕然和何业恒（1980，1981，2006）、何业恒（1994）均认为，历史时期中国孔雀分布的范围很快缩小，数量迅速减少，原因可能是多方面的，其中主要是人为的捕杀，捕杀的主要原因有三：一是食物：早在唐、宋时代，岭南一带即对孔雀大量捕杀，以为食用，如“山谷夷民，烹而食之”；二是装饰用：孔雀羽毛及尾光翠夺目，三国魏文帝曹丕用它作为“车盖”，西汉时南粤用它“珥门户”，古代还用它饰船篷以及扇拂之类；三是药用：宋唐慎微《重修政和本草》卷十九并引《华子诸家本草》提到孔雀可作药用。由于历史时期人们对孔雀无限制的捕杀，加以对山林和草地的不断垦殖，破坏了孔雀的栖息环境。因而使得孔雀的数量迅速减小，分布的范围随着缩小，可见历史时期孔雀分布区的变迁是“人与生物圈”变化的反映，其中人类的活动，特别是对孔雀的大量捕杀，是其变迁中最主要的原因。

王研博和郭风平（2018）认为整个生态系统的内部不对称的互动，也是孔雀种群数量急剧减少，并最终退于云南一隅的原因。事实上，人类、动物和自然作为有机的整体而存在。然而，人类自诞生以来就凭借在自然界中的各种经验而衍生出的技术工具不断地扩展自己的生存空间，这意味着生态系统不断地受到人类活动而产生不确定的影响，而这种影响则不断压缩甚至侵害着自然界中与人类近乎平等的动物。气候、植被、山川，无一不在人类的人口膨胀及生存扩张的进程中遭到摧残，而这些恰恰是孔雀这类具有环境指标功能的动物所赖以生存的条件。

董锋等（2021）研究显示，人类干扰是新石器以来最大的影响因素。

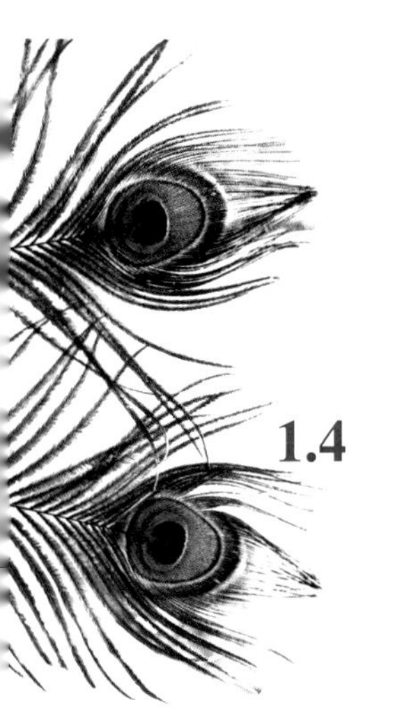

1.4　中国野生绿孔雀分布及数量现状

1.4.1　20世纪我国野生绿孔雀分布调查报道

（1）全国野生绿孔雀分布及数量调查

匡邦郁1960年春夏在云南南部的西双版纳傣族自治州、思茅专区（今普洱市）的墨江县、玉溪专区（今玉溪市）的新平县（即新平彝族傣族自治县）对孔雀生态及捕猎方法进行了初步的观察和访问。根据调查结果认为，国内孔雀仅见于云南南部，尤以思茅（今普洱市）、玉溪（玉溪市）、西双版纳（今西双版纳傣族自治州）、临沧（今临沧市）、德宏（今德宏傣族景颇族自治州）等地最多（匡邦郁，1963）。

文焕然和何业恒（1980）认为，据动物工作者调查，红河哈尼族彝族自治州的蒙自市、河口瑶族自治县、普洱市、西双版纳傣族自治州的勐海县、临沧市、德宏州的盈江县及怒江傈僳族自治州的泸水市一带，现在还有野生绿孔雀分布。

王紫江（1990）报道，于1987年在楚雄紫溪山自然保护区发现孔雀踪迹，1989年1月20日双柏县白竹山自然保护区管理所王兆明曾见到4只（1雄3雌）。1989年2月14日至3月6日在云南楚雄彝族自治州考察时，于2月26日上午10时听到孔雀叫声确认为绿孔雀（*Pavo muticus imperator*），经广泛调查落实，楚雄州禄丰县、双柏县、南华县、姚安县，以及楚雄市的许多地区都有孔雀。

何业恒（1994）认为，云南西南部成为我国野生孔雀仅有的分布地区。目前，我国的孔雀，从现代动物工作者的调查和捕获的标本来看，仅分布在滇西南的红河哈尼族彝族自治州（蒙自市、河口瑶族自治县北）思茅地区（今普洱市）、西双版纳傣族自治州、临沧市、德宏傣族景颇族自治州、怒江傈僳族自治州（泸水市东）一带，与清代比较，虽然泸水市等地是清代文献所不曾记载的，但元江、镇沅、景东、大理、保山等县（市）在清代都产野生孔雀，而现今却已绝灭。可见滇西南野生孔雀的分布范围，又有进一步的缩小。综观上述，可知历史时期中国野生孔雀的地理分布，从北向南，从东北到西南，逐步缩小。

文贤继等通过1991—1993年的信函调查和有选择性的实地访问调查，认为绿孔雀数量较多的地区有云南省的瑞丽市（40~50只）、陇川县（70~90只）、昌宁县（30~40只）、永德县（30~50只）、新平县（40~60只）、普洱市（现为宁洱县，30~40只）、墨江县（30~50只）、景东县（30~40只）、楚雄市（50~80只）、双柏县（150~250只）、南华县（50~100只）、潞西县（今为芒市，数量不详）、龙陵县（10~20只）、云县（15~30只）、临沧（今为临翔区，15~20只）、凤庆县（5~10只）、双江县（数量不详）、沧源县（20~30只）、镇康县（数量不详）、耿马县（数量不详）、景谷县（20~30只）、姚安县（数量不详）、永仁县（数量不详）、禄丰县（数量不详）、维西县（数量需进一步证实）、德钦县（数量需进一步证实）、景洪县（现为景洪市，数量不详）、勐海县（数量不详）、勐腊县（数量不详）、巍山县（数量不详）、盈江县（可能绝迹）、泸水市（可能绝迹）、腾冲县（今为腾冲市，绝迹），并据此认为绿孔雀在中国现仅见于云南西部、中部和南部，过去有分布记录现已绝迹或濒临绝迹的地区有盈江县、泸水市、腾冲市、蒙自市、河口县。永仁县的中和、直左为当次调查发现的分布点，据当地群众反映，在维西县的叶枝，德钦县的拖顶和奔子栏也发现绿孔雀。由于栖息地的破坏，导致绿孔雀现存种群形成小家族群点状隔离分布。各县现存种群的估计数量累加为 635~950只，由于分布数量不详的地区未计入，估计中国现存野生种群数量为800~1100只（文贤继等，1995）。这次调查涵盖了云南省大部分地区，可以认为是中国野生绿孔雀调查史上首次较为全面、系统的一次，虽然给出的是一个估计的种群数量，但对估计中国野生绿孔雀的现状具有重要意义，文章还绘制了20世纪80年代绿孔雀在中国的分布图（图1–2）。

在这篇研究报道中，文贤继等还认为，绿孔雀在历史上曾遍布于湖南、湖北、四川、广东、广西和云南等省（自治区），到20世纪初，其他省区和云南东北部已绝迹，绿孔雀在中国的分布区缩到云南省的西部、中部和南部（文贤继等，1995），这一说法比过去说的“云南省的野生绿孔雀仅分布在滇东北或者滇西南”等的阐述也更为准确，对掌握绿孔雀在云南及中国的分布区更具有指导意义。

杨岚等（1995）以县（市、区）为单位记载了绿孔雀在云南省的地理分布：新平、普洱（今宁洱）、景东、景谷、思茅、墨江、孟连、临沧（今临翔区）、永德、凤庆、沧源、镇康、耿马、弥勒、石屏、蒙自、金平、河口、绿春、勐海、楚雄、禄丰、双柏、南华、姚安、潞西（今芒市）、陇川、盈江、泸水等地。这些分布县（市、区）与文贤继等（1995）报道的基本相同。

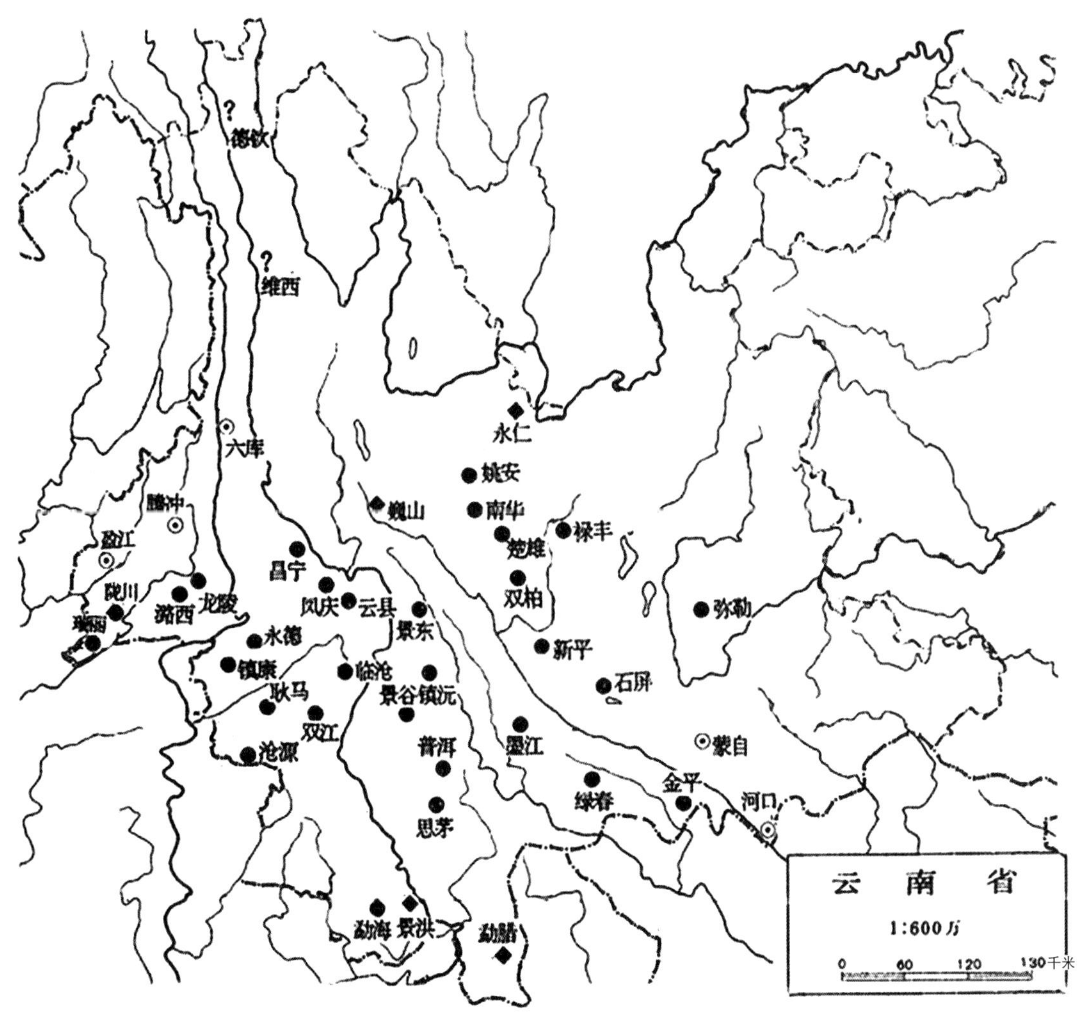

▲ 图1-2　20世纪80年代绿孔雀在中国的分布示意图（文贤继等，1995）

本图境界画法不作划界依据

（2）局部地区野生绿孔雀的分布与数量

①楚雄彝族自治州。徐晖于1990年2月至1995年2月深入云南省楚雄彝族自治州全州绿孔雀有可能分布的县、乡、村，自然保护区及林区，通过亲自观察，调查、访问当地猎人、农民、乡镇干部、自然保护区职工和州内对野生动物有较深研究的前辈，对楚雄州的孔雀分布状况，数量、习性等有较深的了解，根据考察、调查和查阅有关资料，得出了调查结果：绿孔雀在楚雄彝族自治州仅分布于紧靠元江流域以东的楚雄市（包括前进乡、中邑舍乡、大地基乡、吕合乡、东华乡、西舍路乡、朵苴乡、新村乡、云龙乡、母充乡）、双柏县（碍嘉镇、太和江乡、艾尼乡、妥甸镇、雨龙乡）、姚安县（弥兴乡、大河口乡、巴拉乡、左门乡、官屯乡）、南华县（马街乡、徐营乡、 沙桥乡、天申堂乡、雨露乡、龙川镇）、禄丰县（旧庄乡、南河乡、罗川乡）等5个县（市）的29个乡（镇），全州野生绿孔雀数量为280只（其中包括毒死19只、猎杀31只、捕捉23只），其中楚雄市104只、双柏县79只、禄丰县35只、南华县30只、姚安县32只，这5个县（市）有绿孔雀分布的保护区是：楚雄市2个（哀牢山国家级自然保护区楚雄片区、楚雄市紫溪山自然保护区）、双柏县2个（哀牢山国家级自然保护区双柏片区、双柏县白竹山自然保护区）、南华县1个（南华县大中山自然保护区）、禄丰县1个（禄丰县樟木箐自然保护区）（徐晖，1995）。

②临沧市。1992年7—11月，郭宝用等（1999）采用访问调查法和样线法对云南南滚河国家级自然保护区内野生动物进行了调查，遇见6只绿孔雀，估计保护区内有48只绿孔雀。

③西双版纳傣族自治州。罗爱东和董永华于1994年通过访查与建立“听声站” 进行鸣声统计相结合对云南省西双版纳傣族自治州境内野生绿孔雀种群数量及分布现状进行了调查，结果显示，西双版纳现存绿孔雀的种群数量为：19 ~ 25 只（包括在景洪市整糯乡松山林、松毛林一带割松香人所见的 3 只：1雄2雌），仅占中国绿孔雀总数800 ~1100只的2.3%~2.4%；若按西双版纳近 1.969×10^4 km^2 的面积计算，其种群密度仅为（0.95~1.25）$\times 10^{-3}$只/km^2，远远低于全省的平均水平（2.1~2.9）$\times 10^{-3}$只/km^2，已处于濒危状态，亟待保护（罗爱东和董永华，1999）。根据调查资料，罗爱东等还分析了西双版纳地区绿孔雀分布区域缩减率：1980—1985 年，野生绿孔雀的分布面积为 118 km^2；1985—1990 年23 km^2；1990—1995年，其分布面积仅18.7 km^2，与20世纪80 年代相比，分布面积的缩减率为84.15 %；其分布仅局限于景洪市大渡岗乡、景讷乡、勐海县勐海镇西双版纳国家级自然保护区曼稿自然保护区三个狭窄区域内，且现存种群分布区域相互隔离，没

有种群之间的相互交流，严重阻碍了种群的复壮和发展（罗爱东和董永华，1999）。

④云南东南部。杨晓君等（1997）根据1995年、1996年的调查结果认为，云南东南部仅建水县、石屏和弥勒市有绿孔雀分布，与郑宝赉等（1987）和张帆等（1987）记述的分布地点相符。估计云南东南部绿孔雀种群数量50只左右，其中：

建水县：官厅镇和青龙乡20余只；坡头乡不足5只；利民乡有分布，数量不清；盘江乡10~15只。

石屏县：龙朋镇和龙武乡有分布，数量不清；宝秀乡不足 5只。

弥勒市：巡检司，偶见。

而蒙自、金平、绿春、河口、开远和文山等6个县（市）的绿孔雀已经绝迹。另外，个旧市在20世纪80年代初期以前也有绿孔雀，现已绝迹。

⑤ 云南西北部。杨晓君等（1997）根据1995年、1996年的调查结果认为，1986年云南西北部的德钦县支巴洛河边发现2只绿孔雀，为此有关乡村还制定了保护绿孔雀的村规民约，后来1只绿孔雀于1986年3月15日在霞若乡夺松村 （海拔 2200 m）被人捕获，猎捕者因此受到德钦县公安局自然保护区派出所罚款和扣留狩猎工具的处罚，绿孔雀皮被白马雪山自然保护区管理局没收。此后，未再见有绿孔雀分布，该地区出现绿孔雀可能是偶见现象。

1.4.2 21世纪我国野生绿孔雀分布调查报道

（1）全国野生绿孔雀的分布与数量

据文云燕等（2016）报道，2013—2014 年，中国科学院昆明动物研究所调查结果显示，在近5年内，历史分布的 34 个县（市）中仅 11 个县（市）的 14 个地点有绿孔雀野外记录，种群数量估计不足500 只，目前已经成为中国最濒危的野生动物物种之一。

郑光美院士在其主编的《中国鸟类分类与分布名录》（第三版）中，认为绿孔雀在我国分布于西藏东南部、云南（郑光美，2017）。

杨晓君等（2017）报道了2014 年 4 月至 2017 年6月采用问卷、访问和路线法对中国的绿孔雀资源进行调查，同时应用标图、鸣声、红外相机等方法进行补充，并通过查阅文献分析其变化。结果显示，中国有52 个县曾经记录有绿孔雀，但目前仅23个县还有绿孔雀，种群数量已少于 500 只，较文贤继等 1991—1993 年调查的 800~1100 只明显减少。

滑荣等2015—2017年间采用样线法、样点法，结合调查访问进行了绿孔雀种群数量和分布现状调查。选择云南省的玉溪、楚雄、普洱、西双版纳、临沧、保山、德宏、怒江、迪庆、丽江、大理等州（市）的相关地区进行函调和野外实地调查。结果显示，中国现存野生绿孔雀种群数量为 235~280 只，种群数量比 20 多年前 800~1100 只（文贤继等，1995）的调查结果明显减少。分布地区由 1995 年云南省的 32 个县，急剧缩减至现在的 13 个县。确定有绿孔雀分布的地点仅有怒江流域龙陵、永德段局部，澜沧江流域景谷段局部，以及红河流域石羊江、绿汁江沿岸部分地区，共计13个县。并给出了各县（区）的估计数量，分别为：（玉溪市）新平彝族傣族自治县60只；（楚雄彝族自治州）双柏县40~50只；（保山市）龙陵县20~30只；（临沧市）临翔区（未知，待进一步调查）、云县（数量未知，待进一步调查）、凤庆县（数量未知，待进一步调查）、永德县30~40只、镇康县25~30只、耿马傣族佤族自治县（有，数量不详）、沧源佤族自治县（有，数量不详）、双江拉祜族佤族布朗族傣族自治县（有，数量不详）；（普洱市）景谷傣族彝族自治县30只；（德宏傣族景颇族自治州）瑞丽市30~40只。并认为在我国云南，绿孔雀主要生活在沿河谷两岸的热带、亚热带常绿阔叶林和林型较为开阔的低密度思茅松林，由于栖息地的破坏，导致绿孔雀现存种群呈小家族群点状隔离分布（滑荣等，2018）。

Kong等（2018）通过对文献资料查询后，采用实地访问调查法对中国可能有野生绿孔雀分布云南省的11个州（市）和西藏自治区的2个州进行调查，并于2014—2017年间在云南的24个县设置了190条样线（计784 km），在西藏自治区设置了19条样线（计81km），结果表明，在过去的30年间，云南省的11个州（市）中52个县和西藏自治区的1个州（市）的2个县有绿孔雀的分布记录。但20年来，从1991—2000年，分布范围从11个州（市）34个县的127个镇缩减到目前的8个州（市）22个县的33个镇（图1-3），原有分布县的35%，乡（镇）的74%在20年内已经消失了。该研究中，在西藏自治区的墨脱和察隅2个县均未发现绿孔雀。云南楚雄彝族自治州、玉溪市、普洱市、临沧市、保山市、德宏傣族景颇族自治州、红河哈尼族彝族自治州、西双版纳傣族自治州等8个州（市）均位于滇中、滇西和滇南。在云南大理白族自治州、迪庆藏族自治州和怒江傈僳族自治州均没有收集到支持绿孔雀存在的有力证据。在云南中部的元江和峨山地区，发现了以前没有记录的绿孔雀。在这项研究中发现，Kong等发现与20年前相比，绿孔雀的种群数量急剧下降。如，在1991—2000年的调查结果中，云南省有34个县共记录了585~674只，而此次研究采用相同的访谈方法，在云南省的48个县仅记录了183 ~ 240只。在1991—2000年调查记录有绿孔雀分布的34个县中，此次调查结果中，30个县的绿孔雀数量呈下降趋势。在云南省中部双柏县和新平县相邻的两个县绿孔雀数量有明显增长，占云南省孔雀总数量（根据访谈结果）的60%以上（63.93% ~ 69.17%）。

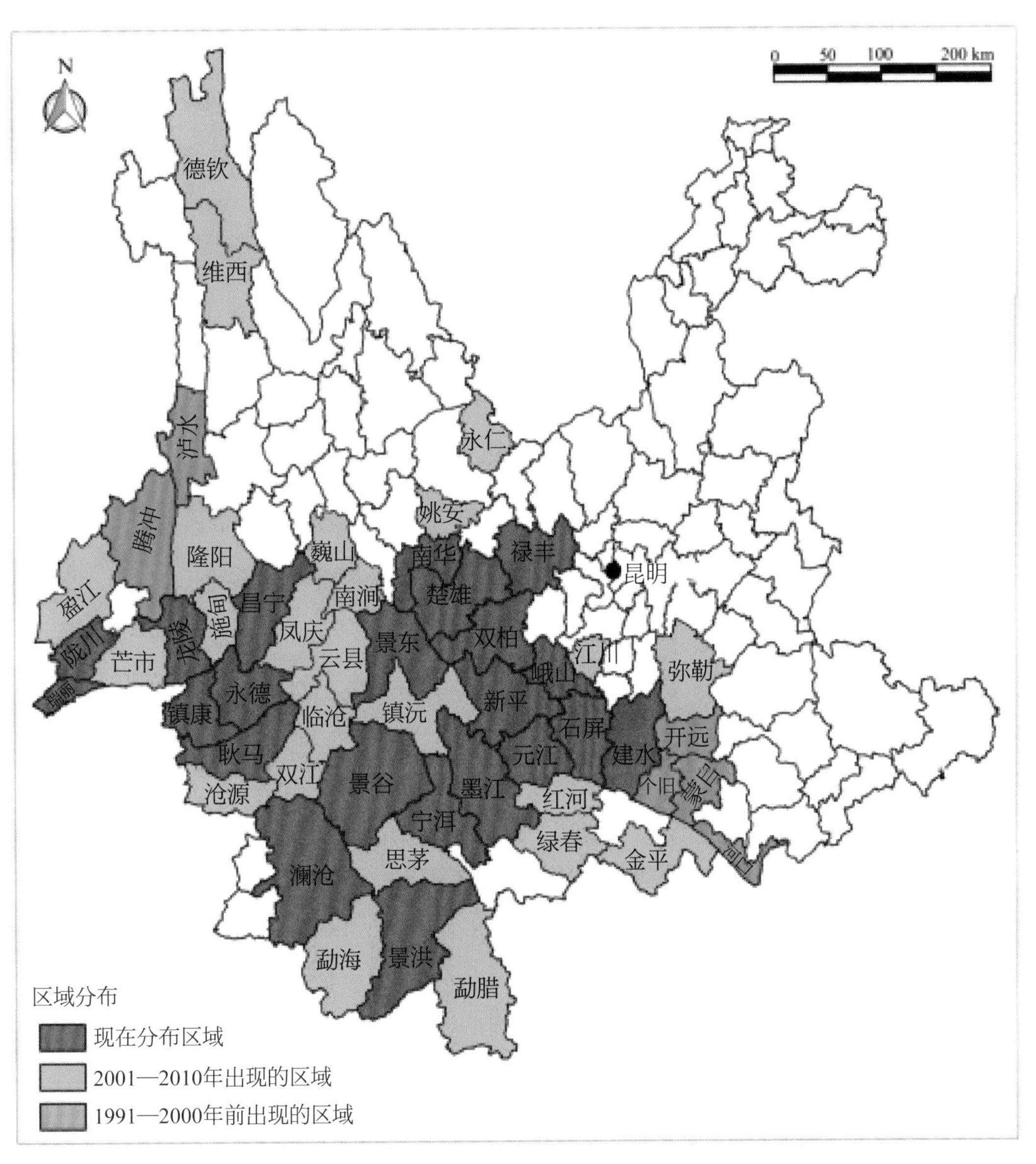

▲ 图1-3　中国绿孔雀分布现状示意图（自Kong et al.，2018，有修改）

本图境界画法不作划界依据

Wu等（2019）提供了中国云南野生绿孔雀分布及种群大小分布图（图1-4），同时也提到，保护绿孔雀所面临的主要挑战是，中国超过65%的绿孔雀生活在保护区外。

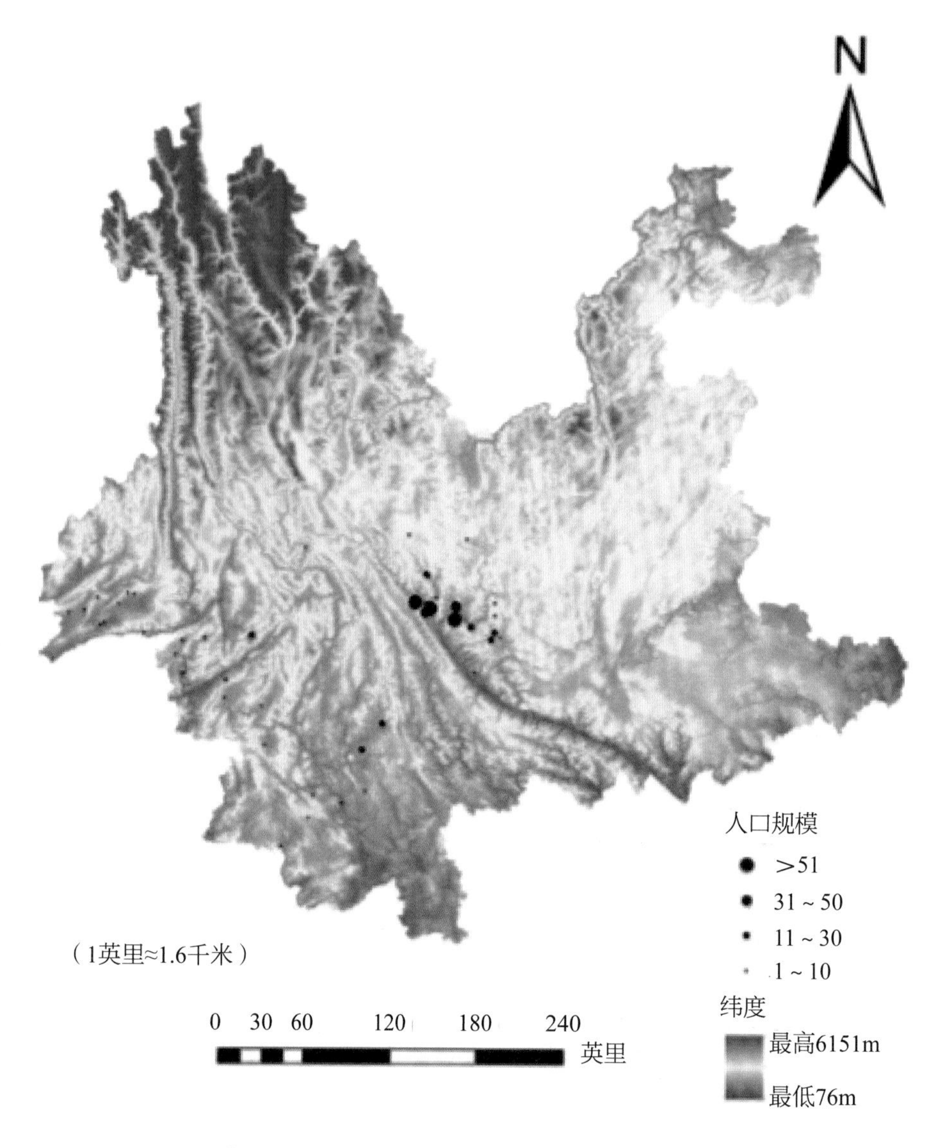

▲ 图1-4 中国云南绿孔雀分布及种群大小分布示意图（Wu et al，2019）

本图境界画法不作划界依据

杨忠兴等（2019）报道，根据2014—2017年中国科学院昆明动物研究所在全国范围内的绿孔雀调查，绿孔雀在中国仅分布于云南中部、西部和南部的 8 个州（市）22 个县（市），估计野外种群数量不足500 只，成为我国目前最为濒危的野生动物物种之一。据 2018年云南省林业和草原局联合中国科学院昆明动物研究所调查，云南省绿孔雀种群数量为485~547只，分布于 6 个州（市）、19 个县。

（2）局部地区野生绿孔雀的分布与数量

①高黎贡山野生绿孔雀分布与数量。艾怀森于1999—2005年间在高黎贡山南段海拔400~1250m的地区记录到了绿孔雀的分布，丰富度为“罕见”（艾怀森，2006）。

② 楚雄双柏恐龙河州级自然保护区绿孔雀分布与数量。文云燕等于2014 年 11 月至 2015 年 7 月间，在双柏县恐龙河州级自然保护区绿孔雀集中分布区域，利用标图法，结合红外触发自动相机（在33个监测点共布设了20台红外相机）开展绿孔雀调查及监测，对获得的绿孔雀图片和视频分析后得出，保护区有绿孔雀成鸟27 只、雏鸟29只（文云燕，2016）。

付昌健等（2019）报道，恐龙河州级自然保护保护区内绿孔雀大约有9群、58只，数量超过巍山青华绿孔雀自然保护区，是中国境内最大的绿孔雀生境保护区。

③ 云南巍山青华绿孔雀省级自然保护区绿孔雀分布与数量。李斌强等2016年10月至2017年9月在云南巍山青华绿孔雀自然保护区的核心区和缓冲区的28个监测位点布设红外相机，累计监测6377台日，共获得独立有效照片1692张，其中兽类563张，鸟类1129张。绿孔雀作为青华保护区的主要保护对象，本次调查并未发现（李斌强等，2018）。

据付昌健等（2019）报道，目前我国以绿孔雀为专门保护动物的保护区仅一个，即云南巍山青华绿孔雀省级自然保护区。该保护区在云南省巍山县青华乡背阴箐、黄家坟、豹子窝一带，距县城47km，其地理坐标为东经100°11′35″~100°14′50″，北纬24°49′45″~25°10′0″。始建于 1988年，1997 年晋升为省级自然保护区。保护区范围1000hm^2。最高海拔 2010.2m，最低漾江边海拔 1146m，立体气候显著，境内有龙凤河和中窑河流经保护区，水资源丰富。保护区内保护有绿孔雀20~30只，由于该保护区面积较小，人类干扰较大，对绿孔雀生存不利，好在保护区的保护力度不断加强，但有关保护区内的绿孔雀种群数量、每只占区面积、保护措施等相关研究及报道未见发表。

④ 瑞丽至孟连高速公路沿线野生绿孔雀分布与数量。李瑞年和林海晏（2018）采用走访调查、红外相机调查和音频监测方法对拟建瑞丽至孟连高速公路途经龙陵和镇康两县，沿线则有 2 处绿孔雀分布点，分别位于龙陵小黑山省级自然保护区的江中山片区和镇康南捧河省级自然保护区的竹瓦片区，数量均已不足20只。

⑤玉溪市新平县野生绿孔雀分布与数量。王方等2017 年1—12 月，采用红外相机对云南省新平县野生绿孔雀开展调查监测，在野生绿孔雀潜在分布区共布设96台红

外相机，累计11482个工作日，在37 个位点拍摄到野生绿孔雀的影像，捕获1370张绿孔雀独立有效照片。根据收集到的数据，结合 3S 技术，作出了绿孔雀分布图。结果显示，新平县野生绿孔雀分布于桂山街道、者竜、老厂、新化、扬武等 5 个乡（镇、街道）的 6 个片区，呈斑块状分布，且这些区域均位于保护区之外，栖息地的破碎化导致各种群间无法交流（图1-5），亟需开展绿孔雀栖息地保护工作（王方等，2018）。

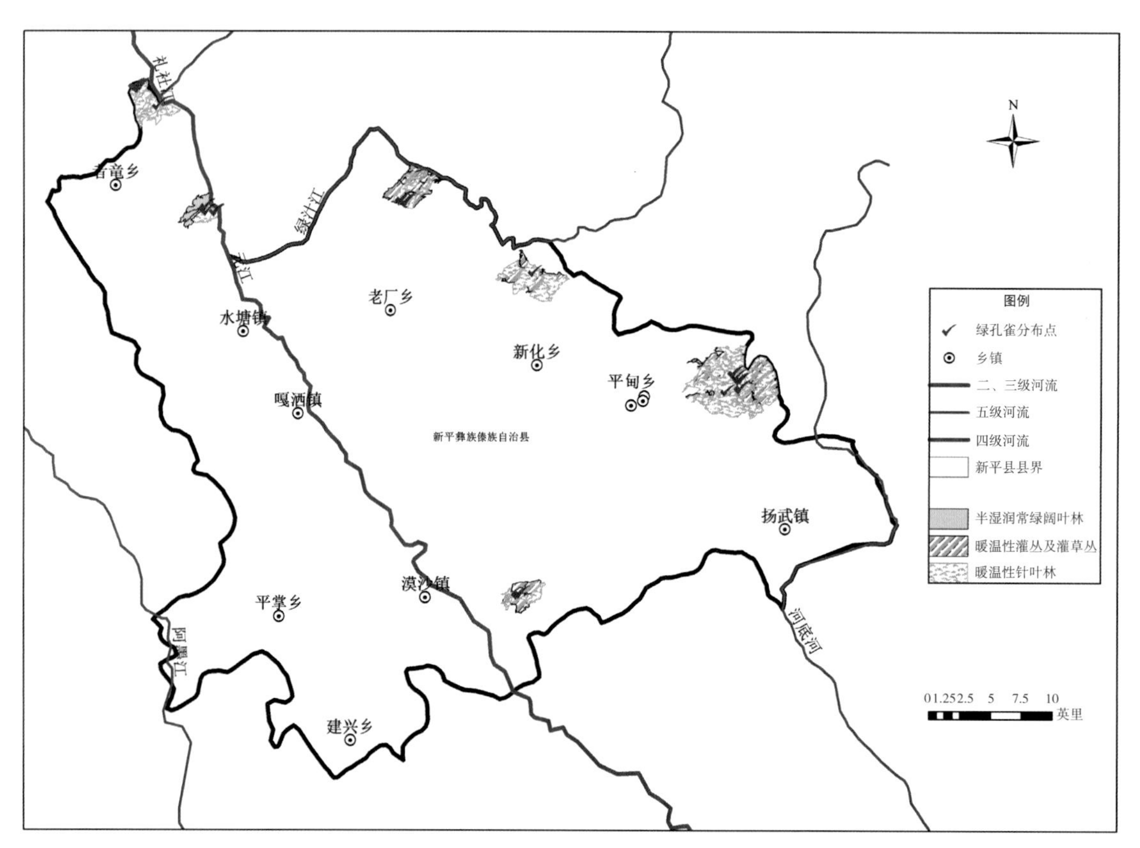

▲ 图1-5　新平县野生绿孔雀分布现状示意图（陈明勇/制）

本图境界画法不作划界依据

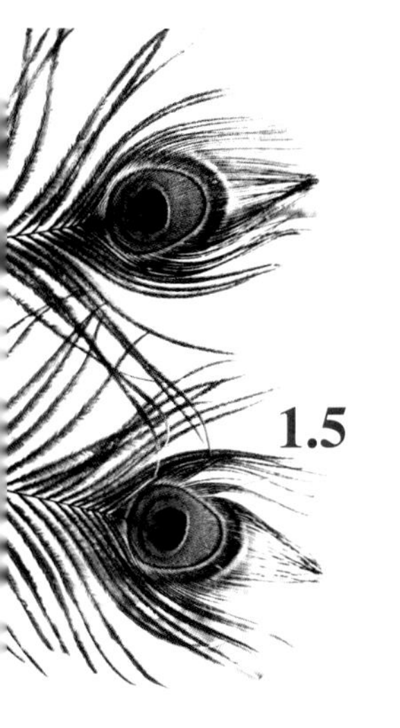

1.5　云南省玉溪市新平县野生绿孔雀种群数量及分布调查

1.5.1　调查研究方法

（1）访问调查法

针对绿孔雀这样色彩艳丽的大型地栖鸟类来说，长期居住在当地的居民对它们更加熟悉和了解，各类人群都很容易识别它们，但凡见过它们的人都会有深刻而清晰的印象。况且，分布区主要是少数民族，民风淳朴，均能如实反映见到的各种情况，信息较准确。因此，采用访问调查法开展野生绿孔雀的分布调查是非常必要的，也是十分高效的。

访问调查步骤：

① 表格编制与完善。首先编制好《野生绿孔雀分布访问调查表》，通过预调查后，对表格进行修改和完善，确保表格中所有字段能全面、准确和具体地反映绿孔雀的分布信息。

② 开展县级调查。对长期在新平县林业和草原局、云南哀牢山国家级自然保护区新平管护局的领导、技术人员、管理人员进行调查，获取绿孔雀分布的初步信息。

③ 乡（镇）级调查。根据县级调查中获取的分布信息，前往相关乡（镇），对乡（镇）领导、林业站、保护区管护所（站）的管理人员、巡护员进行更加详细的访问调查，进一步记录最可能有野生绿孔雀分布的村寨及村小组信息。

④ 村级调查。根据前面的信息，再寻找当地熟悉情况的村小组组长、护林员、放

牧者及普通村民进行访问调查，根据他们提供的分布信息，锁定实地调查的面积、范围及调查研究的方法。

⑤ 对访问调查获得的分布信息进行认真的甄别，确保数据信息的真实、有效。基于对县级和乡（镇）级林业工作人员的访问，对初步确定有或可能有绿孔雀分布的 7 个乡（镇、街道）进行访问调查，对县级访问和乡（镇）访问均确定没有绿孔雀分布的5个乡（镇、街道）不再进行村级调查调查。2017年1—6月，对新平县辖区内7个乡（镇、街道）23 个村（居）民小组共42人进行访问，访问中主要记录受访者是否见过绿孔雀、见到绿孔雀的时间、地点、数量以及受访者的姓名、年龄、职业等信息。为了获得更加确切的分布信息，调查中尤其是对曾经有人见过或听到过绿孔雀叫声的地方增加调查的样本量。

（2）样线法

样线法是在调查区域内选定一定数量的路线开展抽样调查，记录在样线上观察到的动物实体及痕迹相关信息的方法。根据预调查掌握的山形地貌及植被分布情况，调查样线设定为：长3~5 km，单侧宽25 m。调查时段选择在2017年1—6月份，包括部分冬季、全部春季和部分夏季，涵盖鸟类的整个繁殖期。选择天气晴朗的早晨或傍晚进行调查，采用该法沿途观察并记录绿孔雀实体及痕迹（足迹、粪便、羽毛、食痕、卧痕、鸣声等），使用手持GPS、数码相机等设备准确记录发现的时间、地点、生境、植被类型等，有利于分析绿孔雀的生存环境及活动范围。

基于访问调查结果，选择可能有绿孔雀分布的区域布设样线，在新平县老厂乡、者竜乡、新化乡、扬武镇和桂山街道5个乡（镇、街道）共布设10条绿孔雀监测样线，每月沿固定样线调查1次，记录样线调查过程中见到绿孔雀的实体或活动痕迹，2017年1—6月共进行了6次样线调查，累计获得60条样线调查信息。

① 种群密度计算。种群密度是通过绝对数量调查或者取样调查某特定研究地区的个体数量而得到的。根据以上数量调查的结果，可以选用条带最大记数法的密度计算方法来计算绿孔雀种群密度。

条带计数法（许龙等，2003）：

$$D=\frac{n}{2LW}$$

式中，D 为鸟类种群密度；n为样线中绿孔雀数量；L 为样线总长度；W 为单侧样线宽度。

② 种群数量计算。种群数量按公式$N=D \times A$计算（宋志勇等，2018），其中：N为绿孔雀数量，D为种群密度，A为调查区域绿孔雀分布面积，误差为±5%。

（3）红外线自动相机法

① 红外相机布设：红外相机布设点选择在访问调查中受访者见过绿孔雀活动的地方和样线调查中见到绿孔雀实体或活动痕迹的地方，以1km×1km网格在新平县5个乡（镇、街道）（者竜、新化、老厂、扬武、桂山街道）共安装96台自动红外相机，红外相机总覆盖面积为96km^2。根据不同地形条件，红外相机一般安装在树干40~100cm处，机身与树干之间垫上一根木块，使镜头与地面有一定拍摄角度，并保持安装红外相机处的生境原真性，以免对绿孔雀造成影响（王方等，2018）。所使用的红外相机型号有两种：猎科ERE-E1B、猎科Ltl-6210。红外相机拍摄模式设置为“照片+视频”模式，每次独立事件拍摄3张照片和1段10s的视频，照片质量参数为12M，视频画质参数1080P。根据不同环境及采光条件，红外相机感应灵敏度设置为“中”或“低”。红外相机安装时间为2017年1月至2019年6月，分批次每个月陆续收回野外布设的红外相机数据，其中，2017年1月至2017年12月的红外相机数据较完整，因此选取该时段进行数据分析。

② 数据处理方法：一般每个月对红外相机的数据进行一次回收，统一保存到大容量硬盘中，回收数据的同时对红外相机进行检查，确保其能正常拍摄，根据拍摄情况及时更换相机电池和储存卡。

利用Bio Photo红外相机图像处理软件，自动处理提取图像资料的时间信息，对拍摄到绿孔雀的图片和视频按照统一格式进行命名，再人为对拍摄到的物种进行分类鉴定。

1台红外相机安装在野外正常工作24h即记为1个工作日，总有效相机工作日为所有相机工作日的总和。同一台红外相机，每次拍摄到不同的动物则记为1次独立有效事件，如连续拍摄到的是同一动物，则在30min内的所有图像只记为1次独立有效事件（袁景西等，2016）。

③ 个体识别标准：对红外相机拍摄的绿孔雀照片/视频采用个体识别法对进行一一甄别，不同乡（镇）和同一乡（镇）不同区域的绿孔雀归为不同种群，同一区域的野生绿孔雀主要通过个体大小、尾屏长度、羽毛颜色，以及出现在不同红外相机位点的时间等进行个体区分。

1.5.2 新平县野生绿孔雀种群数量及分布调查结果

（1）访问调查结果

对新平县7个乡（镇、街道）19个村（居）民小组42人进行访问，其中有5个乡（镇、街道）15个村（居）民小组34位受访者描述了有野生绿孔雀的分布，平掌乡、漠沙镇2个乡（镇）4个村（居）民组的8位受访者确定无绿孔雀分布（表1–1）。通过对访问数据汇总，得出绿孔雀数量为106~133只，其中，老厂乡为10~20只，桂山街道20~33只，扬武镇7只，者竜乡约66只，新化乡3~7只。

表1–1　绿孔雀访问调查结果统计

访问地点	访问村（居）民小组数（个）	访问人数（人）	访问结果（只）
老厂乡	3	11	10~20
桂山街道	4	5	20~33
平掌乡	3	6	0
扬武镇	4	6	7
者竜乡	3	9	约66
新化乡	1	3	3~7
漠沙镇	1	2	0
合计	19	42	106~133

（2）样线法调查结果

对布设的10条样线共调查6次，累积记录到绿孔雀实体13只/次，发现地点为老厂乡转马都村麻沙丽小组（2只/次）、老厂乡转马都村黑查莫小组他达勒梁子（1只/次）、者竜乡向阳村大象场（1只/次）、者竜乡向阳村中山小组尖山小箐（3只/次）、扬武镇马鹿寨罗韩基地（1只/次）、桂山街道亚尼社区双龙桥小组（4只/次）、新化乡代味村底社柏小组大红箐（1只/次）；此外还听到鸣叫声10处，发现足迹链20条、粪便6堆，收集到雄孔雀繁殖期过后脱落的羽毛53根，并发现1个绿孔雀野外繁殖巢穴和3枚孔雀卵（表1–2），这些踪迹的发现，为红外线相机的布设提供了重要指导作用。

表1-2 样线法调查结果统计

序号	地点	样线长度（km）	实体	羽毛	足迹	鸣叫	粪便	孔雀卵
1	老厂乡转马都村麻沙丽	4.8	2	15	0	0	0	3
2	老厂乡转马都黑查莫小组他达勒	4.2	1	3	0	4	0	0
3	者竜乡向阳村大象场大箐门口	3.5	1	0	0	0	0	0
4	者竜乡向阳村中山小组尖山小箐	3.8	3	7	0	0	0	0
5	者竜乡腰村大春河小组新路沙坝	4.6	0	0	9	3	2	0
6	扬武镇马鹿寨村罗韩基地	4.2	1	4	0	0	0	0
7	桂山街道双龙桥大河山	3.5	1	11	0	0	0	0
8	桂山街道双龙桥斗鸡坡梁子	3.4	2	0	11	0	4	0
9	桂山街道双龙桥小组倒车场	3.9	1	3	0	0	0	0
10	新化乡代味村底社柏小组大红箐	4.5	1	10	0	3	0	0
	合计	40.4	13	53	20	10	6	3

采用样线法6次调查累计记录到绿孔雀13只/次，基于鸟类调查的特殊性，难以通过痕迹来估计鸟类的相对密度，因此都以见到的实体进行计算。分别以见到的绿孔雀数量来计算每条样线的绿孔雀相对密度，得出10条样线绿孔雀的平均密度为0.664只/km^2。

（3）红外相机调查结果

2017年1—12月，对全部96台红外线照相机拍摄的照片和视频影像进行了8次回收，并对图像资料收集整理，结果显示：红外相机累计工作28800个工作日，拍摄到195167张/段照片和视频，其中有21107张/段绿孔雀照片和视频，占全部照片/视频的10.8%。经鉴定，共拍摄到绿孔雀、原鸡、白鹇、野猪等29种动物独立有效照片2309张，绿孔雀独立有效照片1378张。对红外相机拍摄到的绿孔雀图像资料进行一一甄别后，得出红外相机调查拍摄结果（表1-3），共拍摄到126只绿孔雀。

表1-3　红外相机拍摄到各片区野生绿孔雀的数量

乡（镇）	绿孔雀数量（只）	雄性（只）	雌性（只）	幼体（只）
者竜乡	82	19	57	6
老厂乡	9	3	6	0
新化乡	4	0	1	3
桂山街道	26	10	16	0
扬武镇	5	2	3	0
合计	126	34	83	9

（4）新平县绿孔雀种群数量

经访问调查得到绿孔雀数量为106~133只；红外相机调查法拍摄到绿孔雀126只；据样线法调查中的绿孔雀种群密度（0.664只/km^2）和栖息地面积（145.11km^2）可算出，绿孔雀种群数量为91~101只（误差±5%）（表1-4）。红外相机拍摄的结果更精确，因此，确定新平县绿孔雀种群数量为126只。

表1-4　新平县野生绿孔雀数量统计

乡（镇）	布设红外相机数量（台）	访问调查数量（只）	红外相机拍摄数量（只）	样线法结果（只）	综合分析后的数量（只）
者竜乡	47	66	82		
老厂乡	11	10~20	9		
新化乡	6	3~7	4		
桂山街道	12	20~33	26		
扬武镇	20	7	5		
合计	96	106~133	126	91~101	126

1.5.3　关于新平县绿孔雀种群数量及分布的讨论

（1）新平县绿孔雀种群数量、分布及保护地位

本次研究获得了新平县更为全面、翔实的绿孔雀种群数量和分布数据，结果显示：①绿孔雀主要分布于新平县5个乡（镇、街道）的6个片区域（图1-5），种群数量为126只（含9只幼体），较文献记载（文贤继等，1995）数量（40~60只）有了较

大的增长；②者竜乡仍为新平县绿孔雀主要分布点，与之前调查结果相吻合（文贤继等，1995；杨晓君等，2017）；③桂山街道、扬武镇、新化乡为本次调查在新平县发现的绿孔雀新分布点；④调查方法中增加了红外相机调查法和样线调查法，更为全面地获取了新平县绿孔雀的种群现状信息。

从绿孔雀分布图可看出，绿孔雀主要分布于新平县的外围，呈斑块状分布，各区域绿孔雀相距较远，种群之间无法交流，对于绿孔雀数量较小的新化乡和扬武镇，种群面临着巨大的灭绝风险；者竜乡和老厂乡绿孔雀分布于新平县与楚雄市双柏县交界区域，与Han等（2007）的调查结果相吻合。

根据本次调查结果，结合查询到的历次调查数据文献，以及近期云南省林业和草原局、中国科学院昆明动物所等单位组织开展的全省、全国野生绿孔雀种群数量和分布调查的数据，新平县分布的野生绿孔雀及其栖息地具有十分重要的保护价值。

（2）关于野生绿孔雀种群数量及分布调查研究方法

本次调查研究中采用了“访问调查法”“样线法”“红外相机调查法”相结合的调查方法，调查中充分利用了三种方法各自的优势，且方法逐步深入，先通过访问调查获取绿孔雀大致活动区域，再进行样线调查查看绿孔雀活动痕迹，为布设红外相机提供重要参考，最后布设红外相机进行调查，使结果具有较高的可信度。访问调查法获取的信息可以很好地为红外相机布设和样线选择提供线索，也可作为种群数量的一个参考，本次访问结果为106~133只；红外相机的应用使得调查更方便、高效、准确，而且对绿孔雀干扰小，红外相机可以直接拍摄到绿孔雀的照片、视频等资料，可以通过对图像资料的对比识别，从而区分出不同个体，本次红外相机调查得出绿孔雀种群数量为126只；样线法和3S技术的运用，为确定绿孔雀的分布及栖息地范围提供了更准确的方法和数据。

2 野生绿孔雀的形态特征

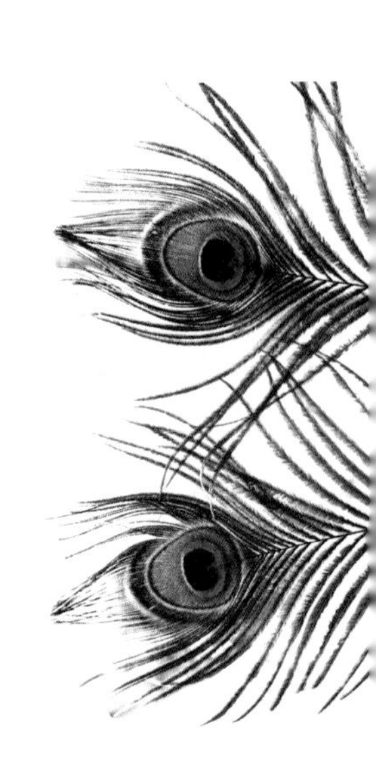

2.1　绿孔雀的分类地位

2.1.1　分类地位

（1）世界孔雀种类

世界上共有3种现存孔雀，即主要分布于东南亚地区的绿孔雀（*Pavo muticus*）（英文名 Green Peafowl）和蓝孔雀（*Pavo cristatus*）（英文名 Blue peafowl），以及分布于非洲刚果盆地的刚果孔雀（*Afropavo congensis*）（英文名 Congo peafowl）（郑光美，2002；孔德军等，2017）。

（2）世界孔雀属鸟类

孔雀属（*Pavo* Linnaeus，1758）全世界只记录到两种，即绿孔雀和蓝孔雀（杨岚等，1995）。

（3）绿孔雀的分类地位

绿孔雀俗名孔雀、诺勇（傣语）、越鸟、南客（见《本草纲目》）（杨岚等，1995；孔德军等，2017），在鸟类分类学中，绿孔雀属于鸡形目（Galliformes）、雉科（Phasianidae）、孔雀属（*Pavo*）、绿孔雀（*Pavo muticus* Linnaeus 1766）。

（4）世界绿孔雀的分布和数量

全世界绿孔雀分布于印度东北部至中国云南、东南亚及爪哇（马敬能等，2000），种群数量在 20 世纪经历了显著的下降，导致分布区的集中和区域种群的灭亡。目前，由于受到干扰和栖息地转变的影响，绿孔雀在东南亚地区仍然面临着种群数量迅速下降的巨大压力，据估计目前全世界的绿孔雀种群数量为10 000~19 999 只成体（孔德军等，2017）。

2.1.2 绿孔雀的亚种分类

（1）绿孔雀的亚种分类

绿孔雀在全世界共记录到3个亚种（杨岚等，1995；孔德军等，2017；滑荣等，2018），分别是：

爪哇绿孔雀亦可称为绿孔雀爪哇亚种（*P. m. muticus*），英文名：Java green peafowl，分布于印度尼西亚的爪哇和马来西亚（滑荣等，2018）；该亚种目前仅分布于爪哇，在马来西亚半岛已经灭绝，在泰国可能也已经灭绝（孔德军等，2017）。

缅甸绿孔雀也可称为绿孔雀缅甸亚种（*P. m. spicifer*），英文名: Burmese green peafowl，分布于印度阿萨姆东南部和缅甸西部（滑荣等，2018），该亚种也可能已经灭绝（孔德军等，2017）。

郑作新（2000）在《中国鸟类种和亚种分类名录大全》中，将*P. m. imperator*的中文名定为绿孔雀滇南亚种（郑作新，2000），而滑荣等（2018）将其中文名称为印度支那绿孔雀，也可称为绿孔雀印支亚种（*P. m. imperator* Delacour 1949），英文名：Indo-Chinese green peafowl，分布于缅甸东部、中国西南地区、泰国和印度支那地区。孔德军等（2017）认为该亚种主要分布于缅甸南部，向东至泰国东部、柬埔寨、老挝和越南，向北一直到中国的南部。

（2）中国绿孔雀的亚种分类

我国学者普遍认为我国野生孔雀仅有一种，即绿孔雀*Pavo muticus*（杨岚等，1995；郑作新，2000；郑光美，2017；孔德军等，2017），并且只有*P. m. imperator*这一个亚种（杨岚等，1995；郑作新，2000；郑光美，2017；孔德军等，2017），但在中文名方面有些差异，有称绿孔雀滇南亚种的（郑作新，2000），也有称印度支那绿孔雀（滑荣等，2018）的，或者也可以称为绿孔雀印支亚种。

2.1.3 绿孔雀的保护级别

绿孔雀被《濒危野生动植物种国际贸易公约》（CITES，1995）列为附录Ⅰ濒危物种（郑光美和王岐山，1998），在中国被列为国家Ⅰ级重点保护野生动物（郑光美和王岐山，1998）。马敬能等（2000）在《中国鸟类野外手册》中描述为“全球性易危（Collar等，994）”。马建章（2002）在《中国野生动物保护实用手册》一书中刊载的经第11次缔约国大会通过，2000年7月19日生效的《濒危野生动植物种国际贸易公约》中，绿孔雀被列入附录Ⅱ中。杨晓君等（2017）报道，绿孔雀在2009年被IUCN从易危种提升为濒危。

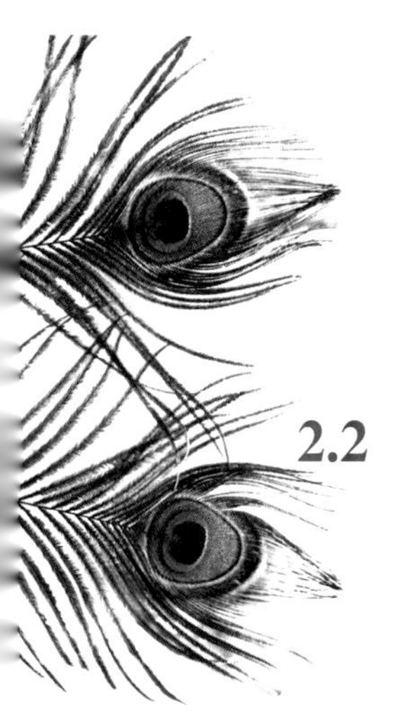

2.2 绿孔雀的形态特征

2.2.1 外形特征

（1）鉴别特征

绿孔雀是我国野生雉类中体型最大的种类，雄鸟体长240cm，成鸟体重可达7700g，翅长415~535 mm（杨岚等，1995）。

成鸟雌雄两性头部特征与雄性相似，体羽主要呈金翠绿色。眼周围具有浅蓝色裸区，颊部裸区呈鲜黄色。嘴峰黑褐色，下嘴较淡；虹膜呈现红褐色。虹膜红褐色，眼周裸出部浅钴蓝色；颊裸出部鲜钴黄色。嘴峰黑褐色，下嘴较淡；跗跖和趾褐色。

成年雄性头上耸立着一簇冠羽，中央部分为辉蓝色，围着翠绿色的宽缘，羽基部和羽干褐色。颈、上背及胸金铜色，羽基暗紫蓝色，常展露于外，羽缘略闪翠绿色；下背和腰翠绿色。具闪耀紫辉的铜钱状花斑；尾上覆羽特别发达，可达100~150枚，长可达1m 以上，羽端具一闪耀蓝色和翠绿色相嵌的眼状斑，形成华丽的尾屏；尾羽黑褐色，形短而隐于尾屏下；初级覆羽和初级飞羽棕黄色。翅面覆盖着黄褐色、青黑色、翠绿色的羽毛，色彩缤纷，华贵艳丽，鲜艳夺目（图2-1，图2-2）；雄鸟跗跖具一长距。雌性成鸟：体形较雄性略小，雌鸟体长110cm，体羽与雄鸟相似，但无尾屏，背羽多呈黑褐色而密布棕褐色斑纹，羽色也较淡钝，远不如雄鸟艳丽。背及腰暗褐色，稍闪黄铜色或绿色金属光泽，并具虫蠹状棕白色波形横纹；内侧覆羽与背相似，但光泽较差；尾上覆羽亦与背相似，但金属光泽较浓；尾羽黑褐色，具褐色横斑和棕白色羽端，并超出尾上覆羽（杨岚等，1995）（图2-3，图2-4）。

幼年雄鸟：与成年雌鸟相似，但眉纹、颏、喉为白色（图2-5，图2-6）。

▲ 图2-1　成年雄性绿孔雀侧面观（云南大学绿孔雀研究团队红外相机/摄）

▲ 图2-2 成年雄性绿孔雀背面观（云南大学绿孔雀研究团队红外相机/摄）

▲ 图2-3　非繁殖期成年雄性绿孔雀侧面观（云南大学绿孔雀研究团队红外相机/摄）

▲ 图2-4 成年雌性绿孔雀背面观（云南大学绿孔雀研究团队红外相机/摄）

▲ 图2-5 繁殖后期的成年雄性绿孔雀背面观（云南大学绿孔雀研究团队红外相机/摄）

▲ 图2-6　绿孔雀幼体（云南大学绿孔雀研究团队红外相机/摄）

（2）绿孔雀局部特征

①头部特征。成年绿孔雀雌雄两性头顶均有一簇别具风度的冠羽长达10cm。冠羽中央部分为辉蓝色，围绕着翠绿色的宽缘，前部呈鱼鳞状，着辉亮的蓝绿色，有时具紫色光泽；脸部，在眼周裸出部分浅钴蓝色，颊皮肤裸露部分鲜钴黄色；嘴峰黑褐色，下嘴较淡（杨岚等，1995）（图2-7）。

▲ 图2-7　成年绿孔雀头部特征（云南大学绿孔雀研究团队红外相机/摄）

②颈部羽色。雄性绿孔雀成鸟颈部和胸部正面（前面）的羽毛金铜色，羽基部暗紫蓝色，羽缘略闪翠绿色；腹部颜色较暗，微微闪着紫光（图2-8）。

▲ 图2-8　雄性绿孔雀颈部羽色（云南大学绿孔雀研究团队红外相机/摄）

③背部羽色。成年雄性绿孔雀背部的羽毛似绿玉一般，周围镶着黑边，中央嵌一个半椭圆形的青铜色的斑，翅膀上覆盖着黄褐、青黑、翠绿的鳞片状羽毛，同样色彩缤纷，在阳光的照耀下，由于羽毛彩色的反光率不同，更显得华丽多彩，鲜艳夺目（图2-9）。

▲ 图2-9 雄性亚成体绿孔雀背部羽色（云南大学绿孔雀研究团队红外相机/摄）

④ 翼及翼上覆羽羽色。成年雄性绿孔雀，初级覆羽和初级飞羽棕黄色，羽端略缀暗褐色；次级飞羽暗褐色，外翈略闪蓝绿色金属光泽；两翅腕缘的覆羽与上背相似，其余覆羽暗蓝绿色或绿蓝色而有金属光泽（图2-10）。

▲ 图2-10 雄性绿孔雀翼及翼上覆羽羽色（云南大学绿孔雀研究团队红外相机/摄）

⑤ 雄性亚成体及非繁殖季节羽色。成年雄性绿孔雀亚成体明显的特征是缺少长的尾羽，其他部位的羽色与成体基本相同。根据绿孔雀的生长发育周期，雄性绿孔雀需要3年以上才能长出华丽的尾屏（图2–11）；非繁殖季节的成体，长的尾羽会自动脱落，在繁殖季节末期，有时会拍摄到一些成年雄性个体背部覆盖有短的末端具眼状斑的尾羽（图2–12）。

▲ 图2–11 尚未长出长尾羽的雄性亚成体绿孔雀（云南大学绿孔雀研究团队红外相机/摄）

▲ 图2–12　繁殖后期长尾羽脱落后的雄性成体绿孔雀（云南大学绿孔雀研究团队红外相机/摄）

⑥成年雌性绿孔雀背部羽色。成年雌鸟绿孔雀羽色不如成年雄鸟绿孔雀艳丽；背羽呈暗褐色，具有黄铜色、绿色光彩，其上密布棕褐色横纹（图2-13）。

▲ 图2-13　成年雌性绿孔雀背羽上暗褐色横纹（云南大学绿孔雀研究团队红外相机/摄）

⑦ 成年雄性绿孔雀尾上覆羽末端的眼状斑。成年雄性绿孔雀在繁殖期（一般在每年的2—6月），尾上覆羽末端有一闪耀蓝色和翠绿色镶嵌的眼状斑，构成一种五色金翠钱纹的图案。最外面是紫色的椭圆圈，次外圈是黄色圈，中间是翠绿的扇形，其上有一蓝黑色的蝶形图案，圈内其余部分为金黄色；圈外还有很多长短不一，呈褐、紫等颜色的细丝，犹如鲜艳夺目的锦缎。

▲ 图2-14　成年雄性绿孔雀尾上覆羽末端的眼状斑（王方/摄）

⑧绿孔雀尾羽及尾下覆羽羽色。成年雄性绿孔雀尾羽短、黑褐色，平时隐于尾屏（尾下覆羽）下，只有在开屏时从后面可看见，与尾屏一起向上翘起，起到一定的支持作用；尾下覆羽暗褐色，似绒状。成年雌性绿孔雀尾羽亦为黑褐色，具褐色横纹的棕白色羽端，并超过尾上覆羽（杨岚等，1995）（图2-15）。

▲ 图2-15　成年雄性绿孔雀开屏时露出的尾羽（云南大学绿孔雀研究团队红外相机/摄）

⑨成年雄性绿孔雀尾上覆羽。成年雄性绿孔雀在繁殖期（一般在每年的2—6月），尾上覆羽特别发达，可达100~150枚，并延长为1m 多长的尾屏，末端有一闪耀蓝色和翠绿色相嵌的眼状斑，闪闪发光（图2-16）。

▲ 图2-16　繁殖期成年雄性绿孔雀尾上覆羽（云南大学绿孔雀研究团队红外相机/摄）

2.2.2　痕迹类别

（1）实体

野外开展野生绿孔雀调查过程中，最好的分布证据是发现绿孔雀的实体：包括成体、幼体和卵（图2-17，图2-18），但在野外环境绿孔雀十分机警，样线调查很难拍摄到好的影像，人工蹲守拍摄成本较大，且会对绿孔雀造成干扰，红外相机拍摄可以在不打扰它们正常生活的前提下拍到好的影像，值得提倡。

▲ 图2-17 用红外线相机拍摄到绿孔雀孵卵的画面（云南大学绿孔雀研究团队红外相机/摄）

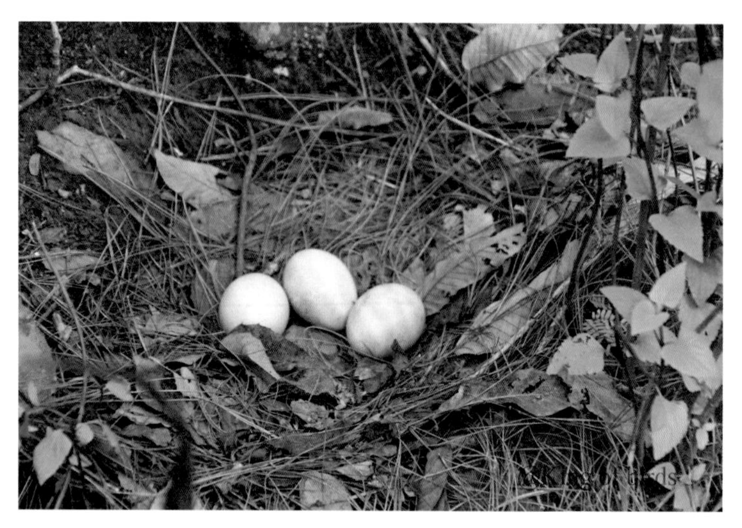

▲ 图2-18　在样线调查中发现的野生绿孔雀的卵（王方/摄）

（2）活动后留下的痕迹

还可以通过一些它们活动后留下的痕迹来初步判断是否有野生绿孔雀的分布，这些痕迹主要有足迹（图2-19～图2-21）、粪便（图2-22）、食痕（图2-23）、扒痕，甚至还会发现它们脱落的羽毛，在调查时要对这些痕迹及空间信息进行测量，表中记录发现的时间、地点（经纬度）、离水源的距离、海拔、坡度、坡向、植被类型、主要植物种类，以及可取食植物的种类，通过这些信息，采用相关的生态模型分析栖息地的适宜性。

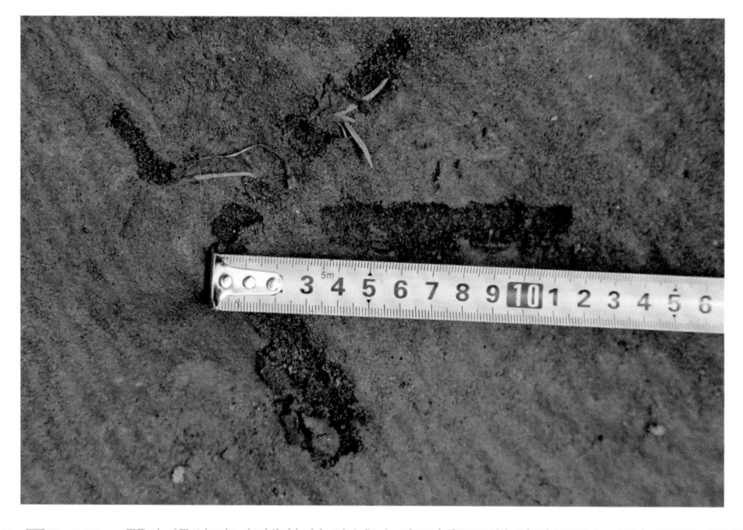

▲ 图2-19　野生绿孔雀在松软的沙滩上走过留下的清晰足迹（陈明勇/摄）

绿孔雀与原鸡、白鹇、家鸡同属于鸡形目、雉科，它们的足迹也很相似，但成年孔雀更大，足迹也更大。

▲ 图2-20 野生绿孔雀在干而松软的沙滩上走过留下的清晰足迹（陈明勇/摄）

▲ 图2-21 野生绿孔雀群体在沙滩上活动留下的足迹和羽毛（王方/摄）

绿孔雀及其他所有鸟类消化系统都较为发达，对食物的消化处理完全，因此它们的粪便中看不到取食的植物或土壤动物的痕迹，粪便也难以用来提取DNA。

▲ 图2-22　在样线调查中发现的野生绿孔雀粪便（王方/摄）

▲ 图2-23　样线调查中发现的绿孔雀啄食痕迹（王方/摄）

（3）羽毛

由于成年雄性绿孔雀在繁殖末期尾上覆羽会脱落的特点（一般在4月底以后就可以发现脱落的羽毛）（图2-24），当然偶然也会发现脱落在地上的其他部位的羽毛（图2-25~图2-28）。在调查中要特别注意寻找样线、样点附近的羽毛，这些是野生绿孔雀分布很重要的凭证。

▲ 图2-24 样线上发现的野生绿孔雀脱落的尾上覆羽（王方/摄）

▲ 图2-25　在样线调查中发现的野生绿孔雀的覆羽（王方/摄）

▲ 图2-26　样线上发现脱落的绿孔雀黄褐色初级飞羽（王方/摄）

▲ 图2-27 样线上发现的绿孔雀翅膀上脱落的青黑色覆羽（王方/摄）

▲ 图2-28 样线上发现的野生绿孔雀脱落的绒羽及覆羽（王方/摄）

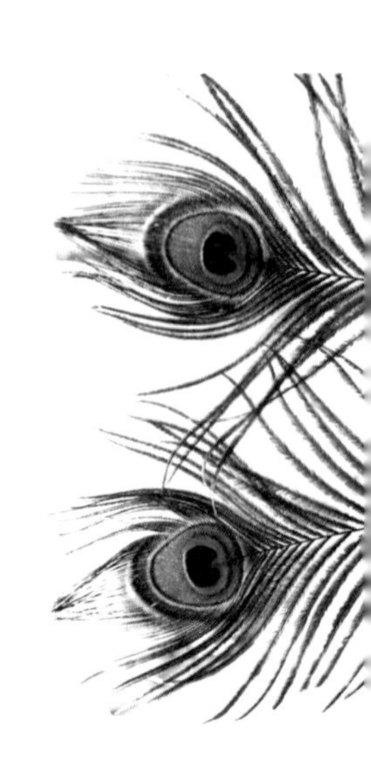

2.3 绿孔雀与蓝孔雀

2.3.1 绿孔雀与蓝孔雀形态比较

蓝孔雀，也叫印度孔雀，是印度的国鸟，原产于印度和斯里兰卡等国。由于蓝孔雀性情温顺易驯化，环境适应能力强，如今人工养殖的数量和规模都很大（聂王星，2017）。目前，我国各动物园饲养的孔雀，绝大多数都是蓝孔雀（图2-29、图2-30）。

▲ 图2-29 雄性蓝孔雀（王方/摄）

▲ 图2–30 雌性蓝孔雀（王方/摄）

蓝孔雀的雌鸟较雄鸟而言，体形较小，体长约1.0m，体重2.7~4.0kg，无尾屏。头部具羽冠，头顶、颈上部呈栗褐色，羽缘带有绿色。眼部、脸部和喉部为白色，颈下部、上背和上胸部为绿色，上体其余部分为土褐色（图2–31）。

▲ 图2–31 昆明动物园饲养的雌性杂交孔雀（王方/摄）

蓝孔雀羽冠为扇形，眼睛上方和下方各有一条白色的斑纹（图2-32）。

▲ 图2-32　杂交孔雀头部的冠羽及跗跖（王方/摄）

绿孔雀颈部呈绿色，其上附着鱼鳞状的羽毛，而蓝孔雀颈部呈蓝色，其上附着丝状羽毛（图2-33）。

▲ 图2-33 杂交孔雀颈部丝状蓝色的羽毛（王方/摄）

绿孔雀羽冠为直立簇状，脸颊为淡黄色。雄性绿孔雀颈部有绿色鳞状羽毛（图2-34）。

▲ 图2-34　成年雄性绿孔雀头、颈部特征（云南大学绿孔雀研究团队红外相机摄）

2.3.2 蓝孔雀的其他色型

（1）白孔雀

白孔雀是在养殖蓝孔雀的过程中，人类选育出的一类可以稳定遗传的变异品种。白化个体由于酪氨酸酶异常，无法合成黑色素，其虹膜颜色很浅，眼睛往往呈现红色。扇状冠羽，全身洁白，羽毛无杂色（图2-35）。

▲ 图2-35 昆明动物园饲养的白孔雀（陈明勇/摄）

（2）杂色孔雀

除白孔雀外，人类还有意识地选育出很多变异品种：尾屏眼斑为白色、翅上覆羽呈蓝黑色、身体羽色蓝白相间的花孔雀等。

2.4 圈养孔雀现状

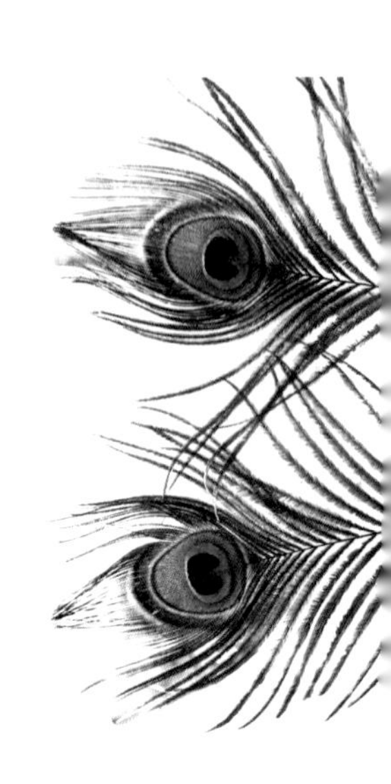

我国近年来提倡发展与旅游业相结合的特色养殖业，孔雀（包括绿孔雀和蓝孔雀）因其具有的观赏价值、经济价值及药用价值，而拥有很大的发展空间。又因孔雀的人工养殖技术简单，适应性强，抗低温，耐高温，因此在我国大多数地区皆可饲养。孔雀养殖环境偏向于野生环境，多适宜半山区的农村养殖。目前，利用农村丰富的饲料资源和地域环境优势，结合当地旅游、餐饮、农副产业协同发展，养殖绿孔雀成了很多农村养殖场的优先选择。

2.4.1 圈养绿孔雀现状

绿孔雀体形大，羽色艳丽，是世界著名的观赏鸟类。曾经在世界上许多动物园中有饲养。但由于野生种群数量急剧下降，在野外获得种源已经十分困难。1989年，绿孔雀在我国被列为国家Ⅰ级重点保护野生动物加以重点保护后，擅自捕捉将受到法律的追究和严厉的处罚。根据我们对国内部分动物园、野生动物园等的调查了解，目前国内动物园内养殖的野生和人工饲养的绿孔雀数量甚少，并存在近亲繁育的问题，也存在绿孔雀与蓝孔雀杂交现象严重等问题，需要开展纯种绿孔雀的培育和种源保护。因此，兼顾保护和利用，开展绿孔雀人工饲养、遗传育种方面的研究，对野生的种群保护及满足观赏方面的需要均很有价值，对这一世界性的珍稀物种保护具有十分重要的意义。

2012年，我们在云南省西双版纳傣族自治州景洪市的曼听公园见到养殖的蓝绿杂交孔雀（图2-36）。

▲ 图2-36 景洪市曼听公园养殖的蓝绿杂交孔雀（陈明勇/摄）

2.4.2 圈养蓝孔雀现状

蓝孔雀是南亚、东南亚、印度次大陆物种。野生蓝孔雀分布于孟加拉国、不丹、印度、尼泊尔、巴基斯坦和斯里兰卡等国家。蓝孔雀是印度的国鸟，也是伊朗的两种国鸟之一。

蓝孔雀分布于东南亚沿海地区，人工驯养繁殖已有 3000 多年的历史，也是世界上著名的观赏鸟类（党心言等，2014）。澳大利亚、巴哈马、新西兰、美国（夏威夷岛）多地有饲养。严晓娟（2009）认为，我国饲养蓝孔雀始于1987年。如果按照王研博等（2018）“蓝孔雀应该才是西域新疆真正的孔雀”的推断，那我国蓝孔雀的饲养历史可能就很久远。国家曾大力鼓励开展蓝孔雀的人工饲养和经营，在全国许多地方都有养殖，包括企业养殖和农户养殖，饲养规模也快速扩大，但具体数据未见报道。

蓝孔雀属鸟纲鸡形目雉科，是世界上著名的观赏鸟类之一（图2-37，图2-38）。它的全身都是宝，其肉可食用，是低脂肪、低胆固醇、高蛋白的保健珍品。1只商品蓝孔雀能卖到1000元左右，而1只体形好的蓝孔雀艺术生态标本，价格大多在10000元左右（唐松元等，2019）。

▲ 图2-37　人工饲养用于观赏的雄性杂交孔雀（陈明勇/摄）

蓝孔雀性成熟期为两年，产卵量可达25~40枚，卵重100~110g，孵化期28d 左右，雌鸟可自己孵卵，育雏母性也很强（朱自强，1997）。

饲养蓝孔雀的目的主要为观赏和食用，食用部分包括肉和蛋，许多地方也用来制作工艺品，包括羽色华丽的雄性蓝孔雀的标本和卵壳标本。在我国，蓝孔雀被作为饲养技术成熟的养殖物种，不属于中国原生物种，可以进行人工饲养、加工和销售。但2020年，由于受新型冠状病毒的影响，全国人民代表大会常务委员会第十六次会议和国家林业和草原局发出通知，禁止食用野生动物，蓝孔雀只可以作为饲养鸟类保留物种，但利用方面仅限于观赏。

▲ 图2-38 人工饲养的正在开屏的雄性杂交孔雀（陈明勇/摄）

3 中国野生绿孔雀的生存环境

3.1 栖息地特征

3.1.1 地形地貌

杨岚等（1995）认为绿孔雀栖息于海拔2000m以下的低山丘陵及河谷地带。徐辉（1995）报道了楚雄彝族自治州分布的野生绿孔雀栖息的海拔高度和坡向随季节而变化。冬春季迁到河谷地带和阳坡，夏季到山中部（最高海拔在1800m）或山箐深处。杨晓君等（1997）报道了1986年云南西北部的德钦县支巴洛河边发现2只绿孔雀，后来1只绿孔雀于1986年3月15日在霞若乡夺松村（海拔2200m）被人捕获。郑光美和王岐山（1998）认为栖息于海拔1250m的低山河谷地带。罗爱东等（1998）报道了云南省西双版纳地区的野生绿孔雀分布的海拔，其中景洪市大渡岗主要活动范围在海拔为1300m左右的区域；勐海县曼稿保护区主要在海拔1300m左右的区域，其次为1250m左右的区域，再次是海拔稍低的平缓山谷边。马敬能等（2000）认为绿孔雀分布高至海拔1500m。艾怀森（2006）认为海拔2500m以下的地域是雉类主要分布区，野生绿孔雀的典型生境为低海拔，并报道了在高黎贡山南段分布的野生绿孔雀海拔在400~1250m之间。孔德军和杨晓君（2017）认为，绿孔雀多栖息于海拔2500m以下的低山丘陵和河谷地带。 Kong等（2018）认为野生绿孔雀分布于海拔2000m以下的区域。付昌健（2019）认为绿孔雀喜在海拔低于2000m的稀树草原，或在生长有灌木、阔叶及针叶等树木的开阔高原地带栖居、活动。

通过对已经发表的上述文献进行分析，我们发现不同的学者对中国野生绿孔雀生活的海拔范围认识不同，总体来看分布范围在400~2500m之间，但尚没见到有关垂直分布方面的专题研究报道。云南大学绿孔雀研究团队在云南省玉溪市新平彝族傣族自治县调查到的野生绿孔雀分布生境为海拔2500m以下的中山、低山及河谷地带（图3-1），植被主要为下坡位低海拔带稀疏的热带季雨林，中坡位的针阔叶混交林及上坡位的针叶林（云南松）带。

▲ 图3-1　元江上游石羊江河谷野生绿孔雀栖息地地貌（陈明勇/摄）

3.1.2　水源

1996年春季，杨晓君等在云南省景东县对春季绿孔雀的栖息地和行为活动进行了初步观察，结果表明，28个样地中所观察的绿孔雀出现在离水源距离小于100m（杨晓君等，2000）。刘钊等（2008）也发现云南元江上游石羊江河谷绿孔雀春、秋季觅食地接近水源，有利于觅食后满足饮水需求。春季绿孔雀多在山间溪流和泉眼等水源附近觅食。小江河河谷地区虽然水源充足，但因地面为砂石基底难以留下足迹，绿孔雀活动痕迹不多，因为两岸地形陡峭，不利于绿孔雀饮水后的扩散。李旭等（2016）对云南楚雄恐龙河保护区绿孔雀春季栖息地选择和空间分布进行研究后认为，该保护区内的绿孔雀觅食地主要选择沿山谷分布的向阳坡面，与对照样地相比坡度较缓，且常接近水源和小路。野外观察也证实，在林区小路上和水源地附近常发现绿孔雀的粪便和足迹。因为动物在觅食过程中总是选择使能量消耗尽可能减小而净收益尽可能获得最大化的觅食对策，上述地形条件有利于绿孔雀觅食后的饮水需求和节省能量。

根据以上文献的分析，水源是绿孔雀生活中的十分重要生态因子，中国野生绿孔雀各分布地绿孔雀对水源这一重要生态因子的选择策略目前还未见有相关专题研究报道。

云南大学绿孔雀研究团队在云南省玉溪市新平县开展调查时，于2017年4月7日在云南元江上游石羊江河谷主河道中，发现了许多绿孔雀活动后留下的清晰足迹（图3-2），表明这里是它们经常活动的场所。调查中还发现，绿孔雀分布区周边的河床除了是绿孔雀主要的水源之外，也是绿孔雀重要的求偶场所（图3-3）。

▲ 图3-2　元江上游石羊江河谷主河道边沙滩上发现的野生绿孔雀足迹（陈明勇/摄）

▲ 图3-3 元江上游石羊江沿岸常因河水冲击形成沙床，也是绿孔雀重要的求偶场所（陈明勇/摄）

3.1.3 植被与植物

（1）中国野生绿孔雀对植被选择方面的研究成果

杨岚等（1995）认为，绿孔雀是热带、亚热带地区的林栖雉类，主要活动于河谷地带的常绿阔叶林及落叶阔叶林、针阔混交林和稀树草地之中，从未见于浓密的热带雨林中。罗爱东等（1998）的研究结果显示，西双版纳地区现存的野生绿孔雀主要在暖性针叶林内（思茅松林）、针阔混交林、亚热带常绿阔叶林、灌木林和种植水稻的平缓山谷边。杨晓君等（2000）根据在云南省景东县对春季绿孔雀的栖息地观察，认为该地区春季绿孔雀的栖息地主要类型为季风常绿阔叶林、思茅松林、针阔混交林、稀树灌丛、荒地灌草丛和农田。刘钊等于2007年的3—4月和10—11月在云南元江上游石羊江河谷绿孔雀的分布区内，采用样线法和样方法调查了绿孔雀的觅食生境，测定了21个生态因子。结果表明，春季的觅食地利用样方距小路距离、乔木种类和藤本密度与对照样方存在显著差异，而秋季的则不显著。生态因子比较和逻辑斯谛回归分

析结果表明，春、秋季绿孔雀均选择落果多、接近水源、坡度小、乔木的盖度和胸径大的地区作为觅食地。乔木和草本盖度，距小路、居民点和林缘距离等是影响绿孔雀春、秋季觅食地选择的关键因子（刘钊等，2008）。李旭等（2016）对云南楚雄恐龙河保护区绿孔雀春季栖息地进行研究后认为，这一区域的绿孔雀多选择乔木高大、郁闭度高、树种和藤本较多的成熟林作为其觅食地，这些区域常常有丰富的落果。郁闭度高的高大乔木表现出绿孔雀选择觅食地时对安全性的要求；丰富的树种和藤本植物以及地面较高的落果密度则体现出觅食地较为丰富的食物种类和数量。

（2）云南省玉溪市新平县野生绿孔雀对植被选择分析

①新平彝族傣族自治县（简称新平县），属云南省玉溪市下辖的自治县，位于云南省中部偏西南，地处哀牢山中段东麓（东经101º16′30″~102º16′50″，北纬23º38′15″~24º26′05″），全县总面积4223 km²，是玉溪市土地面积最大的县。新平县东与峨山彝族自治县相连，东南部与石屏县毗邻，南与元江县山水相连，北隔绿汁江与楚雄彝族自治州双柏县相望（图3-4）。新平县下辖2个街道（桂山街道、古城街道），4镇（扬武、漠沙、嘎洒、水塘），6个乡（新化、老厂、者竜、建兴、平甸、平掌）。

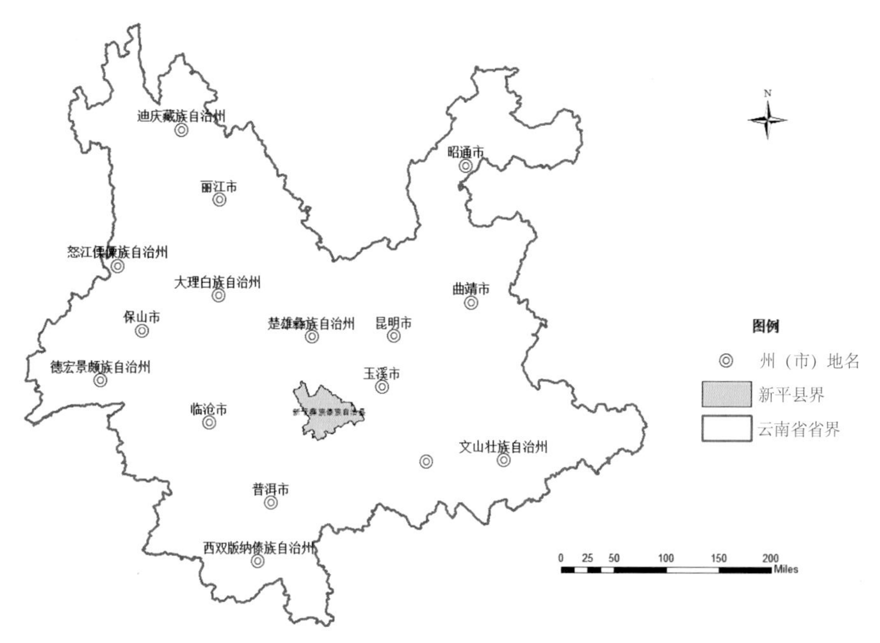

▲ 图3-4　新平县在云南省位置示意图（陈明勇/制）

本图境界画法不作划界依据

② 栖息地分析方法。云南大学绿孔雀研究团队根据云南省2015年卫星影像数据解译，得到新平县植被类型及植被分布图。据访问调查、样线调查及红外相机拍摄结果，判断绿孔雀主要活动区域。由于绿孔雀活动范围内的植被类型为矢量数据，而海拔数据为DEM（栅格）数据，在叠加的过程中，需要对绿孔雀活动区域内的DEM进行重分类，按属性提取2500m以下的区域。 将该栅格图转换为矢量图，再与绿孔雀分布区域植被类型图进行叠加，通过ArcGIS10.2软件计算、分析新平县绿孔雀栖息地类型及面积。

③栖息地分析结果。根据本次调查中植被群落的调查结果，仅在部分植被类型中发现了绿孔雀活动痕迹，如半湿润常绿阔叶林、暖温性灌丛及灌草丛和暖温性针叶林；通过3S技术做出绿孔雀分布图后，结果显示，新平县野生绿孔雀分布于5个乡（镇、街道）共6个片区（者竜乡向阳村、者竜乡腰村、新化乡代味村、老厂乡转马都村、扬武镇马鹿寨村、桂山街道双龙桥村），6片分布区不连接且呈斑块状（图3-5，图3-6）。

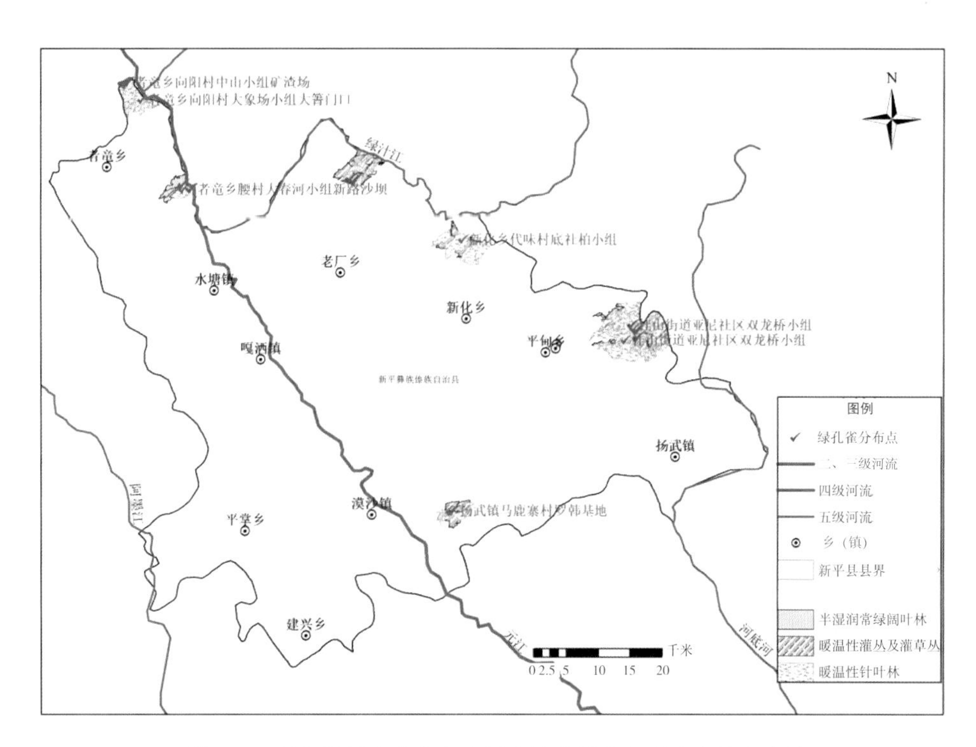

▲ 图3-5 新平县绿孔雀分布及栖息地位置示意图（李正玲/制）

本图境界画法不作划界依据

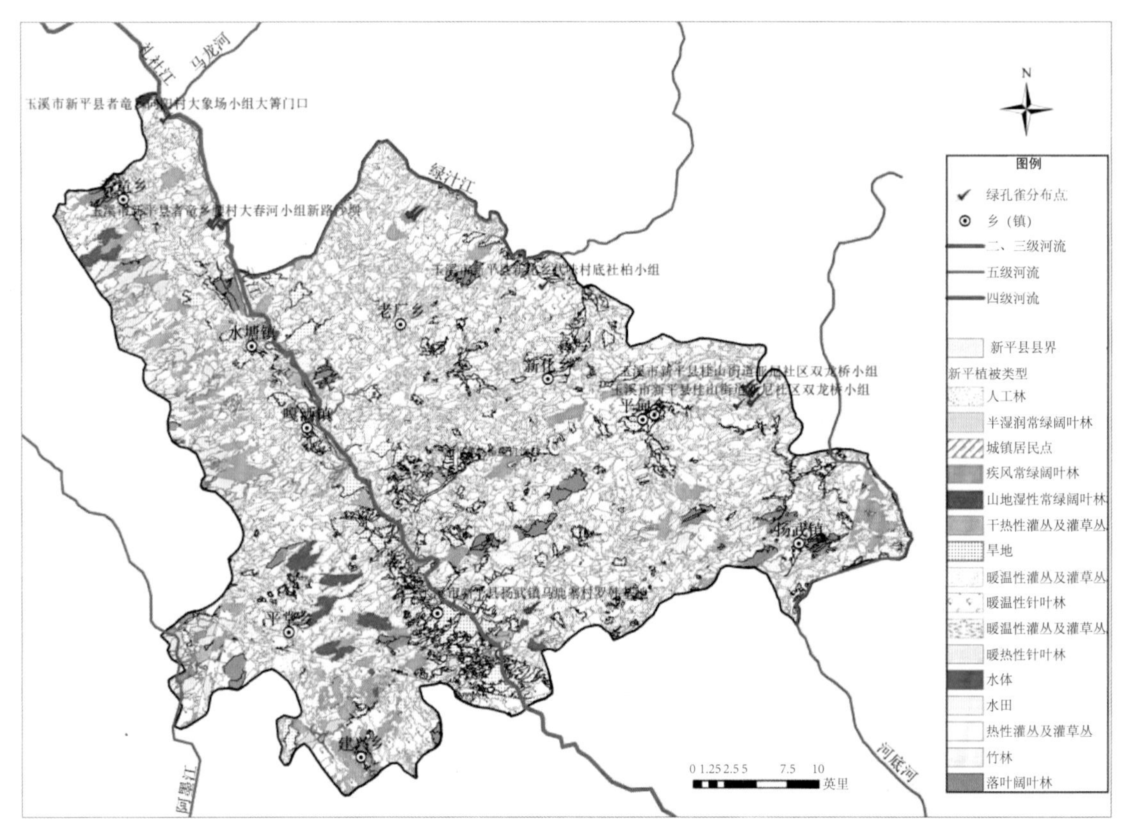

▲ 图3–6　新平县植被及绿孔雀分布位置关系示意图（李正玲/制）

本图境界画法不作划界依据

④ 新平县各片区野生绿孔雀栖息地类型和面积。解译出各栖息地类型和面积，得出全县绿孔雀栖息地总面积为14511hm^2，占新平县总面积（422300hm^2）的3.44%。栖息地类型包括：暖温性针叶林（云南松群落）9759hm^2，占栖息地总面积的67.25%；暖温性灌丛及灌草丛4141hm^2，占栖息地总面积的28.54%；半湿润常绿阔叶林611hm^2，占栖息地总面积的4.21%（表3–1）。

在6片绿孔雀栖息地中，桂山街道面积最大，为6529hm^2，占栖息地总面积44.99%。桂山街道绿孔雀栖息地分两种类型：暖温性针叶林（5112hm^2），暖温性灌丛及灌草丛（1417hm^2）。

老厂乡绿孔雀栖息地类型包括暖温性针叶林（568hm^2）和暖温性灌丛及灌草丛（1339hm^2），总面积1907hm^2。

新化乡绿孔雀栖息地类型包括暖温性针叶林（1712hm^2）和暖温性灌丛及灌草丛

（641hm^2），总面积2353hm^2，暖温性针叶林与暖温性灌草丛交替分布，绿孔雀集中分布在暖温性针叶林中。

扬武镇绿孔雀栖息地类型包括暖温性针叶林（488hm^2）和暖温性灌丛及灌草丛（722hm^2），总面积1210 hm^2。

者竜乡绿孔雀栖息地总面积为2512hm^2，分为两片：向阳村绿孔雀栖息地和腰村绿孔雀栖息地。其中，向阳村绿孔雀栖息地面积为1489hm^2，包括两种类型：暖温性针叶林（1467hm^2）、暖温性灌丛及灌草丛（22hm^2）；腰村绿孔雀栖息地面积为1023hm^2，包括两种类型：暖温性针叶林（412hm^2）、半湿润常绿阔叶林（611hm^2）。

表3-1 新平县各片区野生绿孔雀栖息地类型及面积

新平县绿孔雀分布各片区	栖息地总面积（hm^2）	各类栖息地类型及面积		
		暖温性针叶林（hm^2）	暖温性灌丛及灌草丛（hm^2）	半湿润常绿阔叶林（hm^2）
桂山街道	6529	5112	1417	0
老厂乡	1907	568	1339	0
新化乡	2353	1712	641	0
扬武镇	1210	488	722	0
者竜乡向阳村	1489	1467	22	0
者竜乡腰村	1023	412	0	611
合计	14511	9759	4141	611

⑤ 新平县境内野生绿孔雀对植被的选择偏好的初步分析。新平县境内绿孔雀主要出现在乔灌混交和针阔混交、海拔低于1000m、离水源（流动水源）距离小于200m和土壤较干燥的生境类型中，尤以在麻栎树林和浆果楝林中活动频繁。绿孔雀最喜爱在季雨林中生活，其次为针阔混交林、常绿阔叶林，再次为落叶阔叶林。另据云南大学绿孔雀研究团队此前在新平县各乡（镇）绿孔雀栖息地类型调研结果中获悉，绿孔雀目前仅分布于暖温性针叶林、暖温性灌丛及灌草丛和半湿润常绿阔叶林3种植被类型中，尤其喜欢生活在暖温性针叶林中。

⑥ 新平县境内春季绿孔雀的栖息地主要可分以下4种类型。

a. 常绿阔叶林：以灰毛浆果楝（*Cipadessa baccifera*）林所代表的常绿阔叶林（图3-7，图3-8）。

▲ 图3-7 常绿阔叶林（王方/摄）

▲ 图3-8 绿孔雀生境——稀树灌丛（王方/摄）

b. 季雨林：以小乔木和灌木为主的植被类型多出现在林缘，林冠高通常为5~6m。乔木树种有榕树（*Ficus microcarpa*）、浆果楝和麻栎树（*Quercus acutissima*）等，灌木主要为余甘子 (*Phyllanthus emblica*)、蚂蚱花（*Portulaca grandiflora*）、两面针（*Zanthoxylum nitidum*）、云南松（*Pinus yunnanensis*）、飞机草（*Eupatorium odoratum*）、紫茎泽兰（*Eupatorium adenophora* Spreng）、麻栎树、火绳树 （*Eriolaena spectabilis*）、紫荆木（*Madhuca pasquieri*）、钝叶榕（*Ficus curtipes*）、三叉叶等的季雨林（图3-9）。

▲ 图3-9　绿孔雀生境——季雨林（王方/摄）

c. 针阔混交林：主要为云南松，还与栲类及木荷等混生，形成松、栎混交林。该生境多分布在山坡上，混交林中除云南松(*Pinus yunnanensis*)外，阔叶树有麻栎树、小果栲(*Castanopsis fleuryi*)、余甘子（*Phyllanthus emblica*）、麻栎树、紫茎泽兰、飞机草(*Chromolaene odorata*)、马兜铃 (*Aristolochia debilis*)、蚂蚁菜、老鹳草(*Geranium wilfordii*)等（图3-10，图3-11）。

▲ 图3-10　绿孔雀生境——针阔混交林（王方/摄）

▲ 图3-11　针阔混交林下的绿孔雀（云南大学绿孔雀研究团队红外相机/摄）

d. 落叶阔叶林主要植物有灰毛浆果楝、黄杨木 (*Buxus sinica*)、飞机草、羽叶楸（*Stereospermum colais*）等（图3-12 ~ 图3-15）。

▲ 图3-12　落叶阔叶林下活动的绿孔雀（云南大学绿孔雀研究团队红外相机/摄）

▲ 图3-13　绿孔雀生境——云南松林（王方/摄）

▲ 图3-14　云南松林下活动的绿孔雀（云南大学绿孔雀研究团队红外相机摄）

▲ 图3-15　绿孔雀生境——灌丛（陈明勇/摄）

成年绿孔雀的活动生境也是多种多样，有时会到公路上（图3-16）、林窗空地（图3-17）、林缘（图3-18）、梳林（图3-19）和大树下（图3-20），但前提是这些区域是安全而且没有人为干扰。

▲ 图3-16　路边灌丛中活动的绿孔雀（陈明勇/摄）

▲ 图3-17　针叶林中空地上活动的绿孔雀（云南大学绿孔雀研究团队红外相机/摄）

▲ 图3-18　林缘活动的绿孔雀（云南大学绿孔雀研究团队红外相机/摄）

▲ 图3-19　阔叶林空地上活动的绿孔雀（云南大学绿孔雀研究团队红外相机/摄）

▲ 图3-20 大乔木下活动的绿孔雀（云南大学绿孔雀研究团队红外相机/摄）

3.2 食性

郑作新等（1978）记载：绿孔雀性杂食，嗜吃棠梨、黄泡等果实，也吃稻谷和芽苗、草籽等；此外还兼取食蟋蟀、蚱蜢、小蛾等昆虫，以及蛙类和蜥蜴等。卢汰春等(1991)剖验采自景东林街的1只雌鸟，见有蕈类、幼嫩树叶、草片和白蚁、椿象等，食物总重102g，其中以蕈类食物为最多（杨岚等，1995）。

云南大学绿孔雀研究团队在云南省玉溪市新平彝族自治县开展调查中发现，这里的绿孔雀采食食物种类较为多样化，兼以植物和动物性食物为食，主要包括采食幼嫩多汁的植物叶片、富含水分的浆果，以及昆虫。比如：川梨（*Pyrus pashia*）、黄泡（*Rubus obcordatus*）果实、幼树枝叶、芽苞、蘑菇、草籽、玉米（*Zea mays*）、豌豆（*Pisum sativum*）、稻谷（*Olyza sativa*）等植物和农作物，蚱蜢、蟋蟀、蛾、白蚁、椿象、蚯蚓、蜥蜴、蛙等动物。其中，春季植物性食物有灰毛浆果楝（*Cipadessa cinerascens*）的果实、钝叶榕（*Ficus curtipes*）的果实，以及绿孔雀最喜欢的紫柚木（*Tectona grandis*）果实（图3-21～图3-30）。另外，据当地熟悉绿孔雀习性的护林员齐国发介绍，绿孔雀也会啄食余甘子（*Phyllanthus emblica*）（当地俗称橄榄或滇橄榄）种子（图3-27，图3-28）。野生绿孔雀经常成群去啄食农田间的农作物，特别喜欢豌豆和红薯等农作物。绿孔雀还可通过取食小沙粒来帮助自己消化。

根据文献查询，目前我国野生绿孔雀食性的研究报道还比较零星，还未见关于中国野生绿孔雀的详细的食性植物名录及取食偏好方面的研究报道。为更深入掌握我国野生绿孔雀的生存现状及栖息地质量，对各地分布的野生绿孔雀的食性方面还需要进行深入调查研究。

▲ 图3-21　绿孔雀食物——灰毛浆果楝（*Cipadessa baccifera*）（王方/摄）

▲ 图3-22　绿孔雀食物——聚果榕（*Ficus racenmosa*）（王方/摄）

▲ 图3-23　绿孔雀食物——铁冬青（王方/摄）

▲ 图3-24　绿孔雀食物——虾子花（*Woodfordia fruticosa*）（陈明勇/摄）

▲ 图3-25　绿孔雀食物——水茄（*Solanum torvum*）（王方/摄）

▲ 图3-26　绿孔雀食物——仙人掌（王方/摄）

▲ 图3-27　绿孔雀食物——余甘子(*Phyllanthus emblica*)（王方/摄）

▲ 图3-28　绿孔雀食物——余甘子果实（王方/摄）

▲ 图3-29 绿孔雀食物——虾子花（*Woodfordia fruticosa*）（王方/摄）

▲ 图3-30 绿孔雀食物——紫柚木（*Tectona grandis*）（王方/摄）

4

野生绿孔雀的生态习性

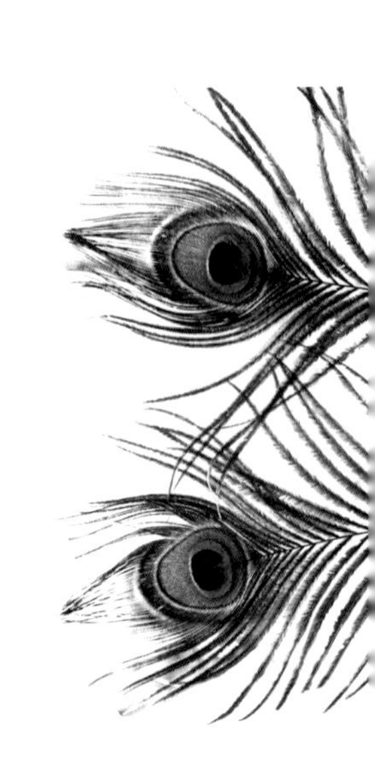

4.1 栖息地选择

4.1.1 觅食地选择

李旭等（2016）对云南省楚雄州恐龙河自然保护区绿孔雀春季的栖息地选择和时空分布研究结果表明，绿孔雀对觅食地和夜宿地的选择与活动行为有关，清晨绿孔雀常从夜宿地所在处下行或滑翔至海拔较低的山谷饮水和觅食，傍晚又向上移动到海拔较高处夜宿，因此绿孔雀有与其他雉类相似的垂直迁徙的日活动规律。无其他干扰的情况下，绿孔雀常在同一区域内觅食或夜宿。食物和隐蔽条件是影响绿孔雀觅食地选择的要素，绿孔雀偏好选择沿山谷分布、坡度较缓的向阳坡面，且常在接近水源和小路的区域觅食，该区域乔木高大、郁闭度高、树种和藤本较多；乔木盖度和胸径是影响绿孔雀夜宿地选择的主要因子，绿孔雀偏好选择乔木郁闭、高大的林型作为夜宿地。绿孔雀的栖息地选择与环境因子之间具有多维相互作用关系，觅食地和夜宿地在环境特征上存在分化，在觅食地和夜宿地之间呈现垂直迁徙的日活动规律。绿孔雀在纬度—海拔方向上的空间分布趋势较为集中，表现为聚集分布在隐蔽条件好且食物和水源丰富的区域（图4-1 ~ 图4-6）。

▲ 图4-1 在疏林中觅食的雄性绿孔雀（云南大学绿孔雀研究团队红外相机/摄）

▲ 图4-2 在箐沟边觅食的雌性绿孔雀（云南大学绿孔雀研究团队红外相机/摄）

▲ 图4-3　在林缘觅食的亚成体雄性绿孔雀（云南大学绿孔雀研究团队红外相机/摄）

▲ 图4-4　在沙砾中觅食的雄性绿孔雀（ 云南大学绿孔雀研究团队红外相机/摄）

▲ 图4-5　常在夜宿地附近觅食的雌性绿孔雀（云南大学绿孔雀研究团队红外相机/摄）

▲ 图4-6　觅食后在夜宿地大树上休息的雌性绿孔雀（云南大学绿孔雀研究团队红外相机/摄）

4.1.2 饮水地选择

根据我们在玉溪市新平县者竜乡的调查，绿孔雀常常在离河流、溪流、沙、水塘不远处活动，表明它们每天活动的过程中与水的关系十分密切，尤其在雨较少的春季，有时它们白天沿着山坡向山上觅食，到了中午或下午时，它们还会寻找水源，补充水分（图4–7）。

▲ 图4–7　发现绿孔雀活动点附近的小溪（陈明勇/摄）

4.1.3　求偶地选择

在每年春季，绿孔雀常选择开阔地集群，成年雄性绿孔雀会展开尾屏，向成年雌性绿孔雀求偶。这种求偶场地的选择主要有林内开阔地（图4-8）、河岸沙滩（图4-9）等。

▲ 图4-8　春季繁殖期间选择林间空地向雌绿孔雀求偶的雄绿孔雀（云南大学绿孔雀研究团队红外相机/摄）

▲ 图4-9　绿孔雀也常常选择无干扰的河边开阔沙滩作为求偶场所（陈明勇/摄）

4.1.4 夜宿地选择

李旭等（2016）的研究结果显示，春季，云南楚雄恐龙河保护区内够绿孔雀选择海拔较高、距水源较远的向阳坡面作为其夜宿地；夜宿地的乔木高大郁闭，树种单一，密度不大，灌木较少，落叶和草本盖度大，藤本和种子密度低。云南大学绿孔雀研究团队在玉溪市新平县开展红外相机监测中也发现了绿孔雀的夜宿地，在附近安装的红外相机拍摄到了上树的绿孔雀（图4-10）。

▲ 图4-10　夜宿地树上停歇的雄性绿孔雀（云南大学绿孔雀研究团队红外相机/摄）

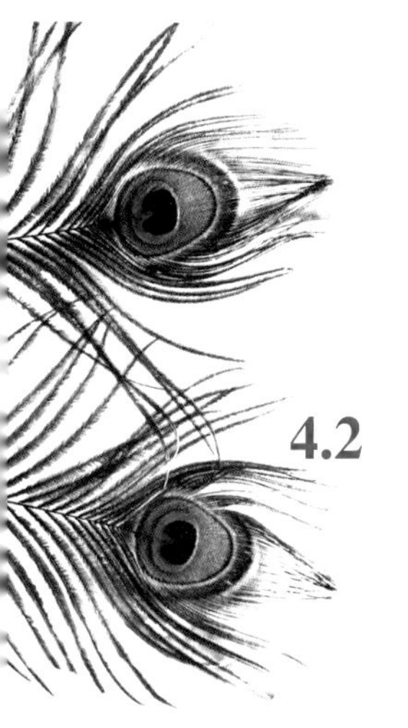

4.2　绿孔雀活动节律

4.2.1　中国绿孔雀日活动节律研究现状

关于中国野生绿孔雀日活动节律方面已经有一些研究报道，如杨晓君等于1996年春季在云南省景东县对春季绿孔雀的栖息地和行为活动进行了初步观察，采用样线法，每天早上7:00—20:00，对观察到的实体和听到的鸣叫声音进行记录，记录23只次绿孔雀的活动主要集中在8:00—13:00和16:00—20:00，取食时间主要为7:00—12:00和18:00—20:00，绿孔雀的行为活动具有上午和傍晚两个较明显的高峰。取食行为占各行为活动时间比例的51.82%，这与笼养绿孔雀的日活动节律和取食节律（杨晓君等，1996）相似。并据此认为，笼养和野外活动节律的相似性可能说明各种动物的日活动节律是在长期进化过程中所形成的（杨晓君等，2000）。李旭等（2016）采用样线法和样方法在云南楚雄恐龙河保护区春季调查时观察到，清晨（6:00—7:00）绿孔雀常从夜宿地所在处下行或滑翔至海拔较低的山谷饮水和觅食，傍晚（17:30—19:30）又向上移动到海拔较高处夜宿。

4.2.2　采用红外相机法对新平县野生绿孔雀日活动节律和年活动节律的分析

2017年1—12月，云南大学绿孔雀研究团队采用红外相机法对云南省玉溪市新平县野生绿孔雀日活动节律分析，结果如下：

（1）研究方法

根据红外相机拍摄的图像资料对绿孔雀活动节律进行分析：

① 日活动节律分析以1h为间隔时段，将每天分为24个取样时段，统计在每个时段各自的独立探测数，并计算各时段的相对活动强度指数（relative activity index，RAI）。

相对活动指数：

$$RAI=\frac{Mg}{M}\times 100$$

其中：Mg表示绿孔雀在g时段的独立照片数，M 表示绿孔雀的总独立照片数。

② 年活动节律分析。动物被红外相机记录到的概率与其活动强度成正相关（李明富等，2011），相对活动指数RAI的大小可以反映动物的活动强度。按月份统计绿孔雀的独立探测数，分别比较各月的RAI大小，分析其年活动节律。

（2）研究结果

① 日活动节律每天各时段活动强度：对2017年1—12月拍摄到的绿孔雀1378张独立有效照片进行分析，并计算各时段（如：6:00—6:59记为“6”）绿孔雀的活动强度（表4–1）。

表4–1　绿孔雀日活动节律

时段	独立有效照片（张）	相对活动强度
6	63	4.57%
7	163	11.83%
8	149	10.81%
9	119	8.64%
10	77	5.59%
11	67	4.86%
12	55	3.99%
13	43	3.12%
14	50	3.63%
15	50	3.63%
16	75	5.44%
17	168	12.19%
18	209	15.17%
19	84	6.10%
20	6	0.44%
合计	1378	100%

利用软件制作的绿孔雀日活动节律图显示：绿孔雀日活动时间为6:00—21:00，其

日活动有明显的高峰和低谷，活动高峰出现在7:00—9:00和17:00—18:00，呈现出双峰模式，7:00和18:00分别至相对活动强度最大值。

12:00—15:00为活动低谷，绿孔雀常在荫蔽的树上或隐蔽处休息；绿孔雀于每日清晨6:00下树后开始觅食，至晚上20:00左右上树休息（图4-11）。

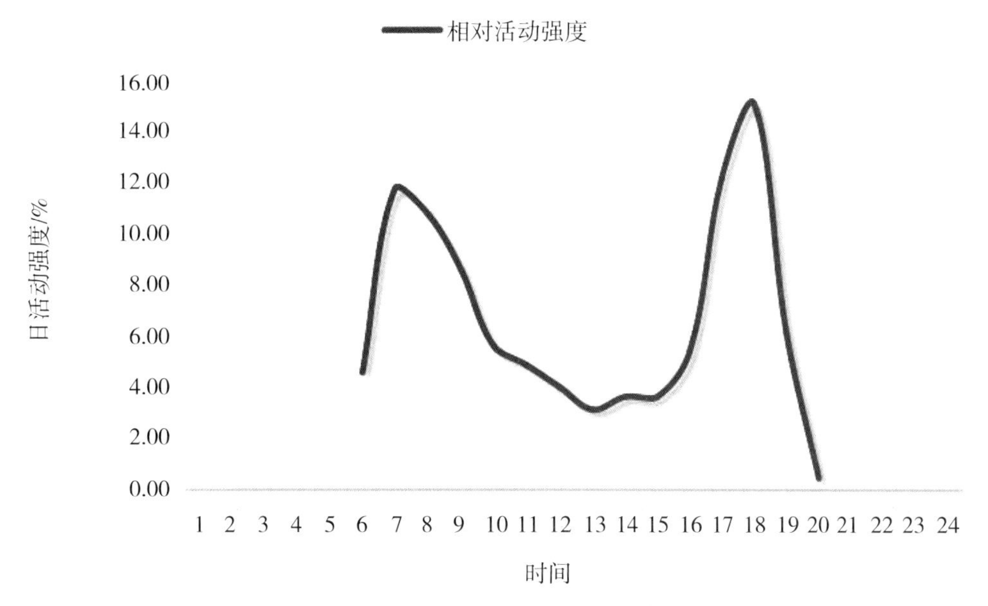

▲ 图4-11　绿孔雀日活动节律图

②关于日活动节律的讨论：通过2017年一整年的红外相机数据分析，我们得到新平县分布的野生绿孔雀的日活动时间为6:00—21:00，与笼养绿孔雀的活动时间一致（杨晓君和杨岚，1996）；与许多雉类活动规律的研究结果类似（赵玉泽等，2013；周晓禹等，2008），比杨晓君在云南景东地区开展的春季野外绿孔雀的活动时间（8:00—20:00）略长（杨晓君等，2000），分析认为，主要原因可能与采用的技术手段和方法有直接关系，红外相机监测比人工开展样线调查和监测更具有优越性，这可能跟红外相机技术应用中，对绿孔雀的活动是24h全天候监测，监测的时间更长有关，包括了天快亮的时段和傍晚天快黑的时段，而这两个时段视线不好，人在山里开展调查和监测是比较困难的，对人身安全也是存在较大的隐患；另外，红外相机是一直处于静态监测中，它们可以在不打扰绿孔雀正常活动的情况下，不失时机地记录下它们活动的时间。因此，采用红外相机开展大型地栖鸟类，尤其是大型雉类日活动节律监测是一种很好的尝试，值得推广应用。

在爪哇岛对绿孔雀的行为生态研究结果显示，绿孔雀的活动时间为5:00—18:00，清晨绿孔雀活动时间和傍晚休息时间均比国内早，推测一方面可能是因为爪哇岛的绿孔雀和国内的绿孔雀属不同的亚种，其生活习性有所差异，另一方面可能是受生活的环境及日出日落时间的差异所影响；爪哇岛绿孔雀在中午会有4~5h的休息，通常会在上午进食和饮水之后，下午进食之前，与国内研究结果相似，这是鸟类躲避阳光直射和休息的一种机制。

a. 7:00—12:00 取食

晨昏是绿孔雀活动的高峰期，清晨5:00—6:00，绿孔雀离开夜宿的高大乔木，滑行至低处或河边，饮水或觅食，野生绿孔雀晨起还会伴有鸣叫，似乎在宣示新的一天开始了（图4-12 ~ 图4-13）。

▲ 图4-12　上午取食的雌性绿孔雀（云南大学绿孔雀研究团队红外相机/摄）

▲ 图4-13 上午取食的雄性绿孔雀（云南大学绿孔雀研究团队红外相机/摄）

b. 12:00—16:00 休息

在中午阳光充足的时候，天气炎热，绿孔雀大多会躲到茂密的灌丛下面，扒一个沙坑：沙浴自洁——将沙子扑到身上进行消毒杀菌，或者趴在那儿梳理羽毛。孔雀的这种行为古人也早已有观察记录 ："喜卧沙中，以沙自浴，拘拘甚适。"（图4-14～图4-15）

经过中午短暂的休息后，绿孔雀又开始下午的觅食活动，直至20:00左右。

▲ 图4-14　下午休息的雌性绿孔雀（云南大学绿孔雀研究团队红外相机/摄）

▲ 图4-15　下午休息的雄性绿孔雀（云南大学绿孔雀研究团队红外相机/摄）

c. 16:00—20:00 取食（图4-16 ~ 图4-17）

▲ 图4-16　日落前取食的雌性绿孔雀（云南大学绿孔雀研究团队红外相机/摄）

▲ 图4-17　下午取食的雄性绿孔雀（云南大学绿孔雀研究团队红外相机/摄）

d. 20:00—7:00 休息

绿孔雀在栖息的时候会保持警惕，一有风吹草动便立刻惊醒飞走。若是在繁殖交配季节，绿孔雀还会发出“ga-wo，ga-wo”的鸣叫声，此起彼伏，响彻整个山谷。若是不识孔雀叫声的人，定会被吓得以为是什么野兽猛禽出没。傍晚来临，绿孔雀会三五成群地回到夜宿地，飞上高大的乔木栖息，绿孔雀对夜栖木的选择也是非常讲究，既要舒适隐蔽，还要高大蔽郁，正所谓“良禽择木而栖”（图4-18）。

▲ 图4-18　傍晚来临前上树休息的雄性绿孔雀（云南大学绿孔雀研究团队红外相机摄）

③ 年活动节律。云南大学绿孔雀研究团队对2017年1—12月拍摄到的绿孔雀1378张独立有效照片进行分析，并计算各月份绿孔雀的活动强度（表4-2）。

表4-2　2017年新平野生绿孔雀年各月份相对活动强度

月份	独立有效照片（张）	相对活动强度
1	43	3.12%
2	343	24.89%
3	333	24.17%
4	262	19.01%
5	73	5.30%

续表4-2

月份	独立有效照片（张）	相对活动强度
6	105	7.62%
7	115	8.35%
8	72	5.22%
9	1	0.07%
10	11	0.80%
11	6	0.44%
12	14	1.02%
合计	1378	100%

年活动节律图显示：绿孔雀2—4月活动强度较大，5月份有所减弱，6—7月活动强度又增加，8月份至次年1月活动强度均较小；绿孔雀年活动节律也出现不明显的双峰模式。据红外相机图像资料显示：2—4月为绿孔雀繁殖前期，开始进入求偶交配季节，5月为雌性绿孔雀产卵、孵卵期，6—8月为绿孔雀育雏期，可见绿孔雀雌鸟带雏鸟外出觅食（图4-19）。

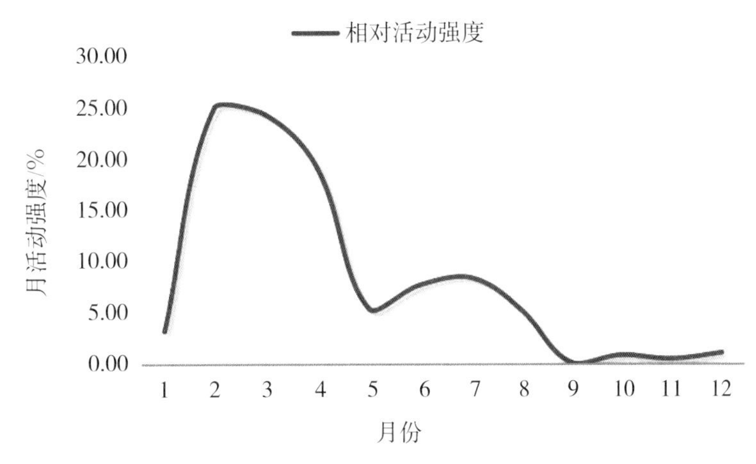

▲ 图4-19 新平野生绿孔雀2017年活动节律图（王方/制）

4.3 绿孔雀与其他物种关系

4.3.1 伴生

（1）绿孔雀与原鸡（*Gallus gallus*）、白鹇（*Lophura nycthemera*）、白腹锦鸡（*Chrysolophus amherstiae*）的日活动节律

绿孔雀与原鸡、白鹇和白腹锦鸡同属于大型地栖雉科鸟类，通过红外相机数据提取四种雉类的独立有效照片和活动时间（表4-3）。结果显示：绿孔雀与原鸡、白鹇、白腹锦鸡均为昼行性雉类，仅在白天活动，其中绿孔雀、原鸡和白鹇日活动节律出现两个活动高峰，白腹锦鸡出现多个日活动高峰（图4-20）。

表4-3　绿孔雀与原鸡、白鹇、白腹锦鸡日活动强度数据

时段	绿孔雀		原鸡		白鹇		白腹锦鸡	
	独立有效照片（张）	相对活动强度	独立有效照片（张）	相对活动强度	独立有效照片（张）	相对活动强度	独立有效照片（张）	相对活动强度
6:00	63	4.57%	11	2.06%	0	0	0	0
7:00	163	11.83%	101	18.88%	5	12.20%	5	12.82%
8:00	149	10.81%	75	14.02%	9	21.95%	4	10.26%
9:00	119	8.64%	29	5.42%	2	4.88%	6	15.38%
10:00	77	5.59%	25	4.67%	0	0	3	7.69%
11:00	67	4.86%	18	3.36%	0	0	4	10.26%
12:00	55	3.99%	18	3.36%	2	4.88%	3	7.69%

续表4-3

时段	绿孔雀		原鸡		白鹇		白腹锦鸡	
	独立有效照片（张）	相对活动强度	独立有效照片（张）	相对活动强度	独立有效照片（张）	相对活动强度	独立有效照片（张）	相对活动强度
13:00	43	3.12%	17	3.18%	0	0	5	12.82%
14:00	50	3.63%	24	4.49%	0	0	0	0
15:00	50	3.63%	24	4.49%	4	9.76%	1	2.56%
16:00	75	5.44%	28	5.23%	3	7.32%	3	7.69%
17:00	168	12.19%	31	5.79%	10	24.39%	3	7.69%
18:00	209	15.17%	99	18.50%	5	12.20%	2	5.13%
19:00	84	6.10%	35	6.54%	0	0	0	0
20:00	6	0.44%	0	0	1	2.44%	0	0

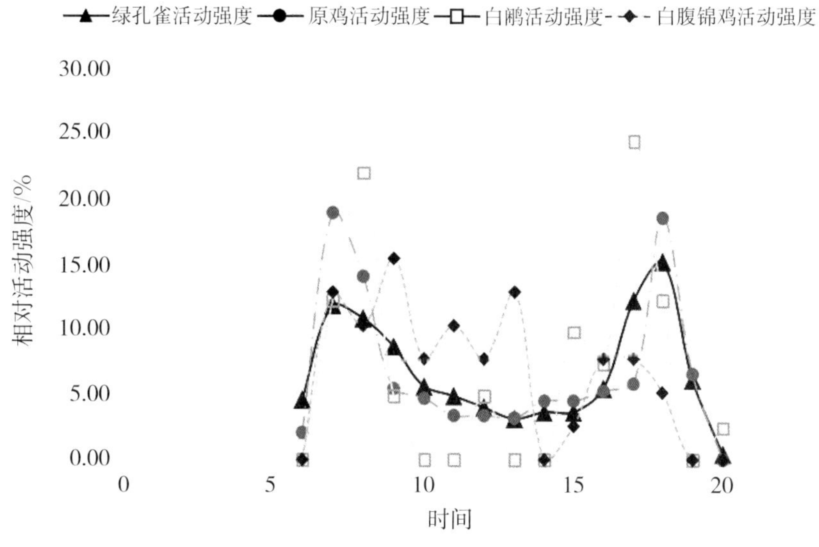

▲ 图4-20 新平绿孔雀与原鸡、白鹇、白腹锦鸡日活动节律（王方/制）

原鸡的日活动时间为6:00—20:00，其日活动有明显的高峰和低谷，活动高峰出现在7:00—9:00和18:00，呈现出双峰模式，7:00和18:00分别至相对活动强度最大值。

白鹇的日活动时间为7:00—21:00，其日活动也有明显的高峰和低谷，活动高峰出现在7:00—9:00和17:00—18:00，呈现出双峰模式，8:00和17:00分别至相对活动强度最大值。

白腹锦鸡的日活动时间为7:00—19:00，7:00—14:00活动强度较大，15:00—19:00活动强度有所减弱，9:00、13:00、16:00、17:00出现活动强度高峰。

绿孔雀与原鸡的日活动高峰出现时间相同，都是7:00和18:00，白鹇的活动高峰为8:00和17:00，白腹锦鸡的活动高峰为9:00，从时间上看，绿孔雀和原鸡的活动高峰时段相同，白鹇上午活动高峰比绿孔雀和原鸡晚1h，而下午的活动高峰比绿孔雀和原鸡早1h；白腹锦鸡上午的活动高峰比绿孔雀和原鸡晚2h，比白鹇晚1h。

对不同时段拍摄的有效照片进行分析统计表明，绿孔雀、原鸡、白鹇、白腹锦鸡的活动均有明显的节律性，绿孔雀、原鸡和白鹇在上午和傍晚各有一个活动高峰，白腹锦鸡在上午、中午和傍晚都出现活动高峰。四种雉类虽为同域分布物种，对资源和栖息地利用相似，属于资源性种间竞争关系，但活动高峰的出现时间有所偏差，绿孔雀上午的活动高峰为7:00，原鸡上午的活动高峰为7:00，白鹇上午的活动高峰为8:00，比绿孔雀和原鸡活动高峰推迟了1h，白腹锦鸡上午的活动高峰为9:00，比绿孔雀和原鸡活动高峰推迟了2h，比白鹇的活动高峰推迟了1h；绿孔雀和原鸡傍晚的活动高峰为18:00，白鹇傍晚的活动高峰为17:00，比绿孔雀和原鸡提前1h，白腹锦鸡中午活动高峰为13:00，傍晚活动高峰为16:00和17:00，与绿孔雀、原鸡、白鹇都有所偏移。活动高峰时间的偏移，使得同域分布且食性和取食方式相似的几种雉类出现了生态位的分离，从而降低其种间竞争的激烈程度。

① 绿孔雀与原鸡一起觅食（图4-21）

▲ 图4-21　雌性绿孔雀与原鸡在一起觅食（云南大学绿孔雀研究团队红外相机/摄）

②绿孔雀伴生种——原鸡（图4-22）

▲ 图4-22　正在觅食的原鸡（♂）（云南大学绿孔雀研究团队红外相机/摄）

③绿孔雀伴生种——白鹇（图4-23～图4-24）

▲ 图4-23　白鹇（1♂1♀）（云南大学绿孔雀研究团队红外相机/摄）

▲ 图4-24　白鹇（♂）（云南大学绿孔雀研究团队红外相机/摄）

原鸡、白鹇、白腹锦鸡食物种类和生活环境与绿孔雀很相似，因此野生绿孔雀与这些鸟类存在食物竞争关系。

④绿孔雀伴生种——白腹锦鸡（图4-25～图4-26）

▲ 图4-25　白腹锦鸡亚成体（云南大学绿孔雀研究团队红外相机/摄）

▲ 图4-26　白腹锦鸡亚成体（云南大学绿孔雀研究团队红外相机/摄）

白腹锦鸡为国家二级保护野生动物，生存环境与绿孔雀较为相似，存在一定的食物及栖息地竞争关系。

成对物种间匹配系数表示绿孔雀与伴生种共同出现的概率，绿孔雀与野猪匹配系数较高，两物种关联性较大。绿孔雀与野猪为非义务性互利共生关系，即：二者因共生而互相获取利益，而二者分开生存时也不会因为彼此间的关系而死亡。从红外相机图像资料及实地观察结果来看，与这一结果是吻合的，野猪刨食树根的同时，把大量的土壤动物也附带刨出，给绿孔雀取食土壤动物带来极大便利，而绿孔雀生性机警，有人为活动或其他危险时，又能给野猪作为警戒哨，从而导致二者的联结性较强，共同出现的概率较大（图4-27 ~ 图4-28）。

▲ 图4-27　绿孔雀伴生种——野猪（云南大学绿孔雀研究团队红外相机/摄）

▲ 图4-28　绿孔雀在野猪刨过的地方觅食（云南大学绿孔雀研究团队红外相机/摄）

点相关系数 *PCC* 指数值可划分 $0 \leq PCC < 0.3$，$0.3 \leq PCC < 0.5$，$0.5 \leq PCC < 0.7$，$PCC \geq 0.7$，$-0.3 \leq PCC < 0$，$-0.5 \leq PCC < -0.3$，$-0.7 \leq PCC < -0.5$，$-1 \leq PCC < -0.7$ 等8个区间来表现种间联结程度，点相关系数（*PCC*）结果显示，绿孔雀与原鸡

（*Gallus gallus*）、蓝喉拟啄木鸟（*Psilopogon asiaticus*）、灰林鸮（*Strix aluco*）、山斑鸠（*Streptopelia orientalis*）、红嘴蓝鹊（*Urocissa erythrorhyncha*）、黑卷尾（*Dicrurus macrocercus*）、赤腹松鼠（*Callosciurus erythraeus*）、鬣羚（*Capricornis sumatraensis*）、赤麂（*Muntiacus muntjak*）均为不显著负联结（$-0.3 \leq PCC < 0$）；绿孔雀与白鹇（*Lophura nycthemera*）、白腹锦鸡（*Chrysolophus amherstiae*）、黑颈长尾雉（*Syrmaticus humiae*）、绿翅金鸠（*Chalcophaps indica*）、紫啸鸫（*Myophonus caeruleus*）、黑领噪鹛（*Garrulax pectoralis*）、猕猴（*Macaca mulatta*）、野猪（*Sus scrofa*）、豹猫（*Prionailurus bengalensis*）、云南兔（*Lepus comus*）为不显著正联结（$0 \leq PCC < 0.3$）。

4.3.2 捕食关系

豹猫（*Prionailurus bengalensis*）、黑熊（*Ursus americanus*）等肉食性动物与绿孔雀之间可能存在捕食和被捕食的关系，成年野生绿孔雀体形较大，机警性强，且移动速度快，所以被捕食的概率很小，但绿孔雀雏鸟经常有被捕食的危险，且野生绿孔雀筑巢简单，一些肉食性动物会偷食孔雀蛋，造成绿孔雀弃巢或减少雏孔雀的出生率（图4–29）。

▲ 图4–29　在绿孔雀分布点附近发现的凤头鹰（云南大学绿孔雀研究团队红外相机/摄）

注：凤头鹰以捕食小型鸟类及爬行类动物为主，会对绿孔雀雏鸟产生威胁

4.3.3 共生关系

赤麂（*Munliacus muntjak*）（图4-30，图4-31）与绿孔雀无明显的食物竞争关系，觅食活动时间也不同，据王方等（2018）研究显示，绿孔雀与赤麂种间关系呈不显著负联结。

▲ 图4-30　在绿孔雀分布点附近发现的赤麂（云南大学绿孔雀研究团队红外相机/摄）

▲ 图4-31　绿孔雀伴生种——赤麂（云南大学绿孔雀研究团队红外相机/摄）

4.3.4 新平野生绿孔雀与伴生鸟兽的种间联结关系分析

（1）种间联结关系研究方法

动物的种间联结关系与物种对生境的选择利用有关，是指不同动物在空间分布上的相互关联性。种间联结度采用2×2列联表（表4-4）计算。将布设的每台红外相机作为一个样方，红外相机拍摄到的动物可记为在样方内出现的动物，以此通过联结表进行计算。本次研究以绿孔雀为目标种，统计绿孔雀与伴生鸟兽在样方内的出现情况，来探讨和分析绿孔雀与其他伴生动物的种间联结关系。

表4-4　2×2列联表

		某伴生物种		Σ
		出现的样方Present	不出现的样方Absent	
绿孔雀	出现的样方Present	*a*	*b*	*a*+*b*
	不出现的样方Absent	*c*	*d*	*c*+*d*
	Σ	*a*+*c*	*b*+*d*	*a*+*b*+*c*+*d*

$$x^2=\frac{N\left[|ad-bc|-\left(\frac{N}{2}\right)\right]^2}{(a+b)(a+c)(b+c)(c+d)}$$

式中：N为取样总数，当$x^2<3.841$时，种间联结性不显著；当$3.841<x^2<6.635$时，种间有显著联结性；当$x^2>6.635$时，种间有极显著的生态联结性。同时可以根据$ad-bc$的正负值来对种间联结性进行判断，$ad-bc$的值大于0时为正联结，$ad-bc$的值小于0时为负联结。正联结表示两物种遇见率和生态位重叠值较高，负联结表示两物种遇见率低，取食空间有差异。

种间联结度计算：

①点相关系数（*PCC*）指数。采用*PCC*系数表示绿孔雀与伴生鸟兽的种间联结程度，*PCC*小于0时表示为负联结，*PCC* 为正值时表现正联结，其绝对值越大，表示联结程度越高，*PCC*为 0 时，则表示种间完全独立。

$$PCC=\frac{ad-bc}{\sqrt{(a+b)(a+c)(b+d)(c+d)}}$$

注：a为都出现的样方数；b为绿孔雀出现而伴生物种不出现的样方数；c 为伴生物种出现而绿孔雀不出现的样方数；d 为绿孔雀、伴生物种均不出现的样方数。

② 成对物种间匹配系数（PC）。选用PC系数表示绿孔雀与伴生鸟兽的种间关联程度和共同出现的概率，在“无关联”时等于 0，在“最大关联”时为 1。

群落系数：

$$PC=\frac{a}{a+b+c}$$

（2）绿孔雀与伴生鸟兽种间联结度

根据红外相机数据，统计出2×2列联表（表4–5），拍摄到绿孔雀与伴生鸟兽的红外相机位点共42个，分别计算绿孔雀与27种伴生鸟兽的种间联结度。

表4–5　绿孔雀与伴生鸟兽种间联结关系

物种	a	b	c	d	x^2	PCC	PC
白腹锦鸡*Chrysolophus amherstiae*	10	18	0	14	4.741	0.395	0.357
白鹇*Lophura nycthemera*	8	20	0	14	3.262	0.343	0.286
原鸡*Gallus gallus*	21	7	1	13	14.616	0.640	0.724
白腰鹊鸲*Copsychus malabaricus*	1	27	0	14	0.128	0.110	0.036
斑背燕尾*Enicurus maculates*	1	27	0	14	0.128	0.110	0.036
大拟啄木鸟*Megalaima virens*	1	27	0	14	0.128	0.110	0.036
黑颈长尾雉*Syrmaticus humiae*	0	28	1	13	0.128	–0.221	0
黑卷尾*Dicrurus macrocercus*	0	28	1	13	0.128	–0.221	0
黑领噪鹛*Garrulax pectoralis*	4	24	1	13	0.028	0.104	0.138
黑胸鸫*Turdus dissimilis*	5	23	0	14	1.391	0.260	0.179
红嘴蓝鹊*Urocissa erythrorhyncha*	1	27	3	11	1.692	–0.287	0.032
灰林鸮*Strix aluco*	0	28	1	13	0.128	–0.221	0
蓝喉拟啄木鸟*Megalaima asiatica*	0	28	1	13	0.128	–0.221	0
绿翅金鸠*Chalcophaps indica*	3	25	0	14	0.404	0.196	0.107
鹊鸲*Copsychus saularis*	2	26	0	14	0.066	0.158	0.071

续表4-5

物种	a	b	c	d	x^2	PCC	PC
山斑鸠*Streptopelia orientalis*	0	28	4	10	5.837	-0.459	0
喜鹊*Pica pica*	1	27	0	14	0.128	0.110	0.036
紫啸鸫*Myophonus caeruleus*	4	24	0	14	0.863	0.229	0.143
豹猫*Prionailurus bengalensis*	14	14	5	9	0.300	0.135	0.424
赤麂*Muntiacus vaginalis*	19	9	5	9	2.734	0.306	0.576
赤腹松鼠*Callosciurus erythraeus*	7	21	0	14	2.593	0.316	0.250
果子狸*Paguma larvata*	1	27	1	13	0.066	-0.079	0.034
猕猴*Macaca mulatta*	4	24	0	14	0.863	0.229	0.143
明纹花松鼠*Tamiops macclellandii*	2	26	0	14	0.066	0.158	0.071
野猪*Sus scrofa*	10	18	5	9	0.117	0	0.303
云南兔*Lepus comus*	1	27	1	13	0.066	-0.079	0.034
帚尾豪猪*Atherurus macrourus*	9	19	0	14	4.771	0.369	0.321

注：x 为种间联结性指数；PCC为点相关系数；PC为成对物种匹配系数

种间联结指数x^2结果可分为三个等级：$x^2>6.635$时，种间关联极显著；$x^2<3.841$时，种间关联性不显著；$3.841<x^2<6.635$时，种间关联显著。绿孔雀与伴生鸟兽种间联结验证结果显示：

①绿孔雀与原鸡（14.616）有极显著的生态联结性（$x^2\geq6.635$）。

②绿孔雀与山斑鸠（5.837）、帚尾豪猪（4.771）、白腹锦鸡（4.741）种间生态联结性显著（$3.841<x^2<6.635$）。

③绿孔雀与白鹇（3.262）、赤麂（2.734）、赤腹松鼠（2.593）、红嘴蓝鹊（1.692）、黑胸鸫（1.391）、紫啸鸫（0.863）、猕猴（0.863）、绿翅金鸠（0.404）、豹猫（0.3）、白腰鹊鸲（0.128）、斑背燕尾（0.128）、大拟啄木鸟（0.128）、黑颈长尾雉（0.128）、黑卷尾（0.128）、灰林䳭（0.128）、蓝喉拟啄木鸟（0.128）、喜鹊（0.128）、野猪（0.117）、鹊鸲（0.066）、果子狸（0.066）、明纹花松鼠（0.066）、云南兔（0.066）、黑领噪鹛（0.028）种间联结性不显著（$x^2<3.841$）。

4.3.5 点相关系数

点相关系数*PCC* 指数值可划分 0≤*PCC*＜0.3，0.3≤*PCC*＜0.5，0.5≤*PCC*＜0.7，*PCC*≥0.7，–0.3≤*PCC*＜0，–0.5≤*PCC*＜–0.3，–0.7≤*PCC*＜–0.5，–1≤*PCC*＜–0.7共8个区间来表现种间联结程度。

点相关系数（*PCC*）结果显示：

①绿孔雀与帚尾豪猪（0.64）、原鸡（0.395）、黑领噪鹛（0.369）、山斑鸠（0.343）、喜鹊（0.316）、蓝喉拟啄木鸟（0.306）、紫啸鸫（0.26）、黑卷尾（0.229）、鹊鸲（0.229）、白腰鹊鸲（0.196）、斑背燕尾（0.158）、果子狸（0.158）、灰林鵖（0.135）、白腹锦鸡（0.11）、白鹇（0.11）、赤麂（0.11）、黑颈长尾雉（0.11）、黑胸鸫（0.104）均为正联结，其中，与帚尾豪猪关联性最强。

②绿孔雀与明纹花松鼠（0）无关联性。

③绿孔雀与野猪（–0.079）、云南兔（–0.079）、赤腹松鼠（–0.221）、红嘴蓝鹊（–0.221）、绿翅金鸠（–0.221）、豹猫（–0.221）、猕猴（–0.287）、大拟啄木鸟（–0.459）均为负联结，其中与大拟啄木鸟关联性最强。

4.3.6 成对物种间匹配系数

成对物种间匹配系数*PC*数值可划分为 *PC*＜0.3，0.3≤*PC*＜0.5，0.5≤*PC*＜0.7，*PC*≥0.7，4个区间来表现种间联结程度。

成对物种匹配系数（*PC*）结果显示：绿孔雀与黑领噪鹛（0.724）的关联性最强，相遇的概率最大；其次分别是野猪（0.576）、明纹花松鼠（0.424）、帚尾豪猪（0.357）、大拟啄木鸟（0.321）、豹猫（0.303）、原鸡（0.286）、云南兔（0.250）、白腰鹊鸲（0.179）、黑胸鸫（0.143）、红嘴蓝鹊（0.143）、鹊鸲（0.138）、白腹锦鸡（0.107）、白鹇（0.071）、绿翅金鸠（0.071）、山斑鸠（0.036）、喜鹊（0.036）、蓝喉拟啄木鸟（0.036）、黑颈长尾雉（0.036）、赤腹松鼠（0.034）、猕猴（0.034）、斑背燕尾（0.032），与紫啸鸫（0）、黑卷尾（0）、果子狸（0）、灰林鵖（0）、赤麂（0）无关联性。

4.3.7 关于绿孔雀与伴生鸟兽种间关系的讨论

种间联结指数x^2表示两物种的遇见概率和生态位重叠率，显著或极显著正相关的物种对，趋向于生活在相似的空间内，或具有相似的栖息环境需求或生态位。通过绿孔雀与伴生的27种鸟兽进行种间联结关系验证表明，绿孔雀与原鸡有极显著的生态联结性，表明绿孔雀与原鸡生活空间和对栖息地选择利用相似，遇见率和生态重叠率较高，研究结果与绿孔雀和原鸡日活动节律结果相吻合。绿孔雀与山斑鸠、白腹锦鸡和帚尾豪猪的生态联结性显著，遇见率高；与其他鸟兽生态联结性不强，遇见率低。

点相关系数（*PCC*）结果显示：绿孔雀与帚尾豪猪、原鸡、黑领噪鹛、山斑鸠、喜鹊、蓝喉拟啄木鸟、紫啸鸫、黑卷尾、鹊鸲、白腰鹊鸲、斑背燕尾、果子狸、灰林䳭、白腹锦鸡、白鹇、赤麂、黑颈长尾雉、黑胸鸫均为正联结，其联结性依次减弱，与帚尾豪猪关联性最强，与明纹花松鼠无关联性；绿孔雀与野猪、云南兔、赤腹松鼠、红嘴蓝鹊、绿翅金鸠、豹猫、猕猴、大拟啄木鸟均为负联结，其联结性依次增加，其中与大拟啄木鸟关联性最强。

成对物种匹配系数（*PC*）反应两物种共同出现的概率，其值域为[0, 1]，数值越接近1，表明其关联程度越紧密。成对物种匹配结果显示，绿孔雀与黑领噪鹛关联性最紧密，其次分别是野猪、明纹花松鼠、帚尾豪猪、大拟啄木鸟、豹猫、原鸡、云南兔、白腰鹊鸲、黑胸鸫、红嘴蓝鹊、鹊鸲、白腹锦鸡、白鹇、绿翅金鸠、山斑鸠、喜鹊、蓝喉拟啄木鸟、黑颈长尾雉、赤腹松鼠、猕猴、斑背燕尾，绿孔雀与紫啸鸫、黑卷尾、果子狸、灰林䳭、赤麂无关联性。

造成正相关主要有两个原因：一是一个物种依赖另一个物种；二是在异质的环境内，几个物种对环境条件有相似的适应和反应；造成负联结主要是因为二者在资源竞争中相互排斥。与绿孔雀成正联结关系的物种中，没有依赖绿孔雀或者绿孔雀依赖其生存的物种，因此与绿孔雀成正联结关系的物种是在异质环境中对环境条件产生相似的适应和反应造成的；与绿孔雀成负联结关系的物种，是二者在资源竞争中产生排斥作用造成的。

5 野生绿孔雀的行为特征

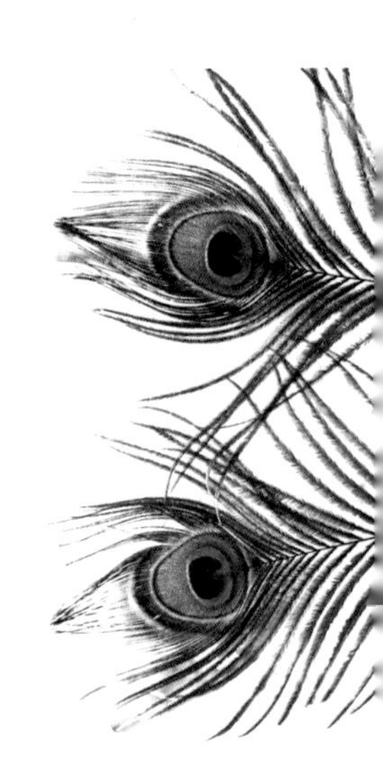

5.1　觅食行为

绿孔雀觅食活动是用喙直接啄食地上掉落的植物果实和种子，对不太高的植株，也会啄食其细嫩的树叶，还会用喙、爪挖掘草根和腐败的落叶下的昆虫（图5-1～图5-3）。绿孔雀一天的时间分配中，取食所占时间比例最多（51.82%），取食高峰主要为7:00—12:00和18:00—20:00两个时段（杨晓君等，2000）。

▲ 图5-1　绿孔雀寻找食物（云南大学绿孔雀研究团队红外相机/摄）

▲ 图5-2　正在获取食物的绿孔雀（云南大学绿孔雀研究团队红外相机/摄）

▲ 图5-3　绿孔雀寻找食物（云南大学绿孔雀研究团队红外相机/摄）

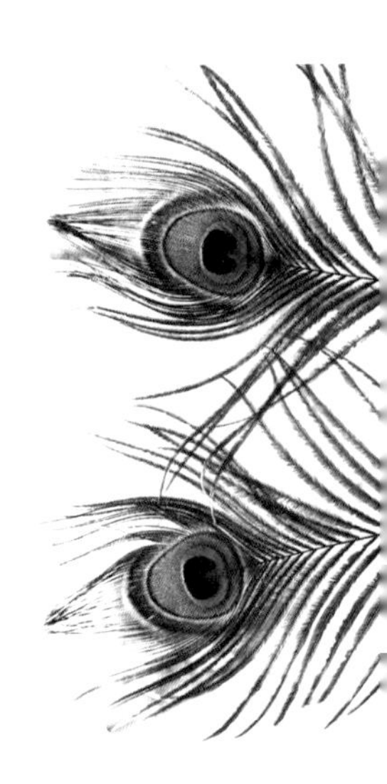

5.2 社群行为

绿孔雀为典型的一雄多雌鸟类，通常结成5~10只群体活动，每个群体中只有1只成年雄鸟，其余为雌鸟或亚成体组成的家族群（图5-4 ~ 图5-7）。活动群体的大小随季节而有所变动，一般在冬天会集成大群，其余季节呈分散的家族群活动。

▲ 图5-4 绿孔雀雄鸟和雌鸟一起觅食（云南大学绿孔雀研究团队红外相机/摄）

偶，有时会十分激烈。

▲ 图5-5 进食中的绿孔雀（云南大学绿孔雀研究团队红外相机/摄）

▲ 图5-6 雌性绿孔雀常结伴觅食（云南大学绿孔雀研究团队红外相机/摄）

成年雄性绿孔雀勇猛好斗，繁殖期内雄性间的争斗主要是通过展开尾屏炫耀求偶，有时会十分激烈。

▲ 图5-7　雌雄绿孔雀结伴觅食（云南大学绿孔雀研究团队/供图）

每年2月中下旬，在云南哀牢山下石羊江两岸，绿孔雀进入繁殖季，此时河谷及两旁山中会响起雄性绿孔雀求偶的叫声，相和而鸣，此起彼伏，响彻漫山——这是雄性绿孔雀在宣示主权。若有其他绿孔雀在自己的地盘内鸣叫，它便会靠近入侵者，鸣叫警告，驱逐其离开，若对方不离开，双方可能会展开一场激烈的争夺配偶的战斗。

一个小的绿孔雀家族群只有1只雄性绿孔雀成体，主要负责保卫整个家族群其他孔雀安全以及种群繁衍，不同家族群雄性孔雀之间都有各自的家域（图5-8）。

▲ 图5-8　单独觅食的雄性绿孔雀亚成体（云南大学绿孔雀研究团队红外相机/摄）

绿孔雀群体除了一起觅食外，在休息时会各自梳理羽毛（图5-9）。

▲ 图5-9　聚在一起梳理羽毛的雌性绿孔雀群体（云南大学绿孔雀研究团队/供图）

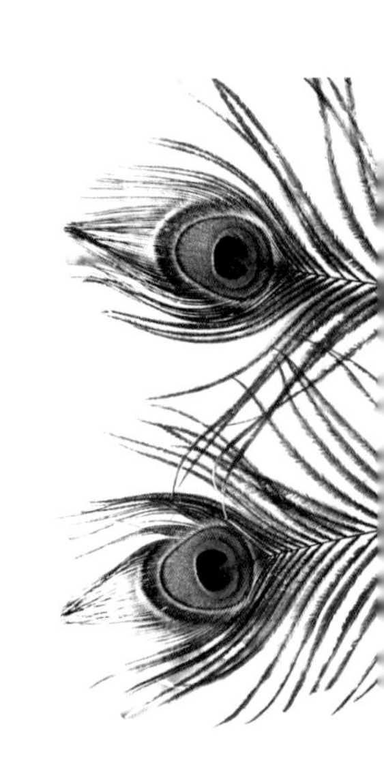

5.3 繁殖行为

绿孔雀的繁殖期一般为每年的3—6月，发情的雄鸟有优美的求偶炫耀行为，它们会激动地将尾屏高举展开，支撑在翘起的尾羽上，形如大扇子，左右摇摆和转动，当它们快速抖动尾屏时，美丽的眼状斑闪闪发亮，异常高雅、优美和华丽，这种精彩的求偶表演，就是通称的 “孔雀开屏”（图5-10～图5-11）。开屏次数0～15次，开屏时间长短不一，短的几秒钟、长的达两小时之久。 开屏后雄孔雀通常通过身体的转动向雌孔雀“夸示”，左右旋转，翎羽上的眼状斑反射着光彩。有雌孔雀在雄孔雀面前站立或经过，雄孔雀会间断性地快速高频率地抖动尾屏发出“沙沙”声，并时常跺脚引起雌孔雀注意。孔雀交配行为是一种行为链，其过程是雄孔雀求偶，雌孔雀应答（趴卧）、爬跨，啄咬雌孔雀头颅、踩踏、交尾退下。交尾成功后雄孔雀不会立刻收起尾屏，会继续进行交尾。

▲ 图5-10 正在开屏的成年雄性绿孔雀（云南大学绿孔雀研究团队红外相机/摄）

▲ 图5-11 繁殖期的绿孔雀群体（云南大学绿孔雀研究团队红外相机/摄）

孵卵由雌鸟单独承担，孵化期为27~30d，啄壳至完全出雏平均需要30h。雏鸟出壳前1~2d可听到壳内“吱吱”的叫声和啄壳声，叫声由弱到强，数小时后，啄壳声逐渐增强，频率随之增加。雏鸟在距离卵钝端1~1.5cm横径处啄出一小孔，然后顺着卵壳亚中部啄出一个将卵分成大头端部分与小头端部分的裂缝，最后将壳脱掉，完全出雏（图5-12）。

▲ 图5-12　雌性绿孔雀正在孵卵（云南大学绿孔雀研究红外相机/摄）

求偶配对成功后，于3—4月份进入筑巢孵卵期。绿孔雀通常营巢于隐蔽性较好的灌草丛中，为地面巢，于地面凹陷处覆以树枝、杂草等，极其简陋。绿孔雀每窝产卵3~6枚，卵为椭圆形，呈乳白色或乳黄色，光滑而无点斑（图5-13）。

▲ 图5-13 绿孔雀巢及卵（王方/摄）

雏鸟属早成性，出壳后即可跟随雌鸟四处觅食，雄性幼鸟需3年后方可长出华丽的尾屏（图5-14）。

▲ 图5-14 雌性绿孔雀育雏（云南大学绿孔雀研究团队红外相机/摄）

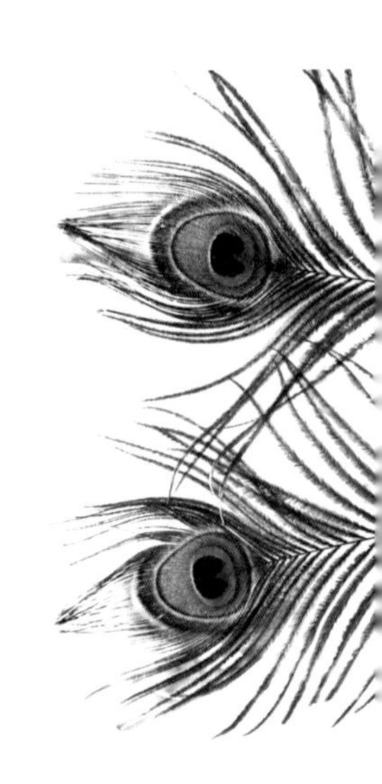

5.4 躲避天敌

孔雀性机警，脚强健，善于奔跑，不善于飞翔，但下落时速度较快；喜欢群居生活，很少单独活动，秋冬群集更大。野生绿孔雀遇到天敌时，会大声发出鸣叫并快速奔走或短距离滑翔以躲避天敌追捕，大多数时候会依靠那双大长腿迅速逃遁（图5–15）。

▲ 图5–15　正在四处张望的雄性绿孔雀（云南大学绿孔雀研究团队红外相机/摄）

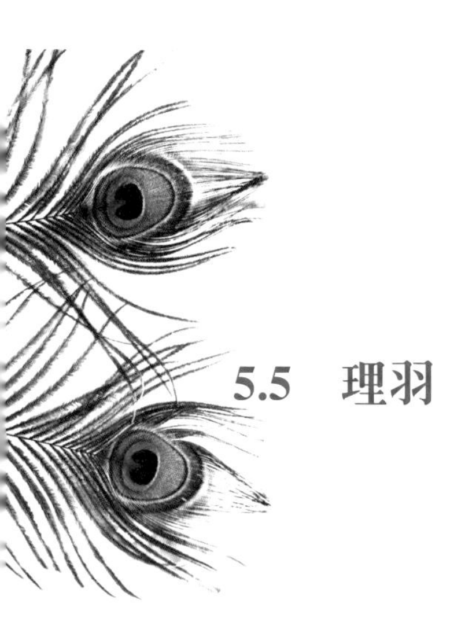

5.5 理羽

理羽是绿孔雀日常生活的重要活动之一，当它们休息时会选择安全的地方用喙梳理自己的羽毛，将尾脂腺上的油脂涂抹在羽毛上，保持羽毛的光泽，雌、雄成体都特别注重理羽，尤其是在繁殖期的雄性成年绿孔雀（图5-16～图5-17）。

▲ 图5-16 绿孔雀梳理羽毛（1）（云南大学绿孔雀研究团队红外相机/摄）

▲ 图5-17　绿孔雀梳理羽毛（2）（云南大学绿孔雀研究团队红外相机/摄）

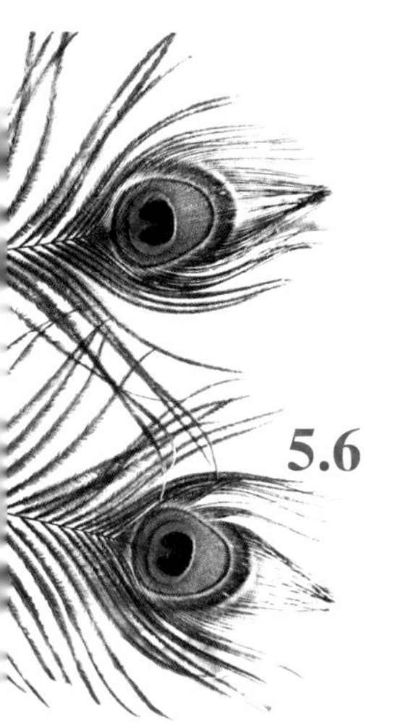

5.6　开屏、行走、观察和警戒行为

野生绿孔雀的行为多种多样，由于它们性情十分机警，野外观测中仅能靠红外相机捕捉到的照片和视频获得行为特征片段，有待长期观测积累更多的数据进行深入的分析，才能更全面了解它们野外的行为和生活。绿孔雀的开屏、行走、观察及警戒如图5-18~图5-21。

▲ 图5-18　孔雀开屏（云南大学绿孔雀研究团队红外相机/摄）

▲ 图5-19　孔雀行走（云南大学绿孔雀研究团队红外相机/摄）

▲ 图5-20　繁殖后期雄性绿孔雀脱掉华丽的尾上覆羽（云南大学绿孔雀研究团队红外相机/摄）

▲ 图5-21 独自活动的绿孔雀警戒状态（云南大学绿孔雀研究团队红外相机/摄）

6 中国野生绿孔雀的研究及保护现状

中共云南省委、云南省人民政府、云南省林业和草原局、云南省生态环境厅对绿孔雀保护管理做了大量工作，取得了一定成效。第一是划建保护区，在云南建立的各级自然保护区中，有15处保护区有绿孔雀分布，并将绿孔雀作为主要保护对象之一进行保护管理。第二是将绿孔雀列入极小种群物种，2007 年云南省林业厅（2018年改为云南省林业和草原局）率先在全国提出极小种群物种保护，并将绿孔雀作为优先保护的20个重点物种之一。第三是开展种群数量调查与监测，全国第一次、第二次野生动物调查中调查了绿孔雀； 2017 年、2018 年云南省林业和草原局联合中科院昆明动物研究所分别启动了“元江上游绿孔雀种群和栖息地调查与评估”和“全省绿孔雀资源调查”工作，进一步查清了云南省绿孔雀资源现状，为科学保护绿孔雀奠定了基础。第四是建立绿孔雀栖息地共管保护小区，由云南省林业和草原局牵头，阿拉善资助，在专家指导和职能部门的支持下，2018 年实施了“新平县腰村绿孔雀栖息地共管保护小区”项目，项目实施取得了良好效果，为探索绿孔雀保护管理模式奠定了基础（图6–1 ~ 图6–2）。

▲ 图6–1 开展绿孔雀分布调查（王方/摄）

▲ 图6-2 开展绿孔雀栖息地调查（陈明勇/摄）

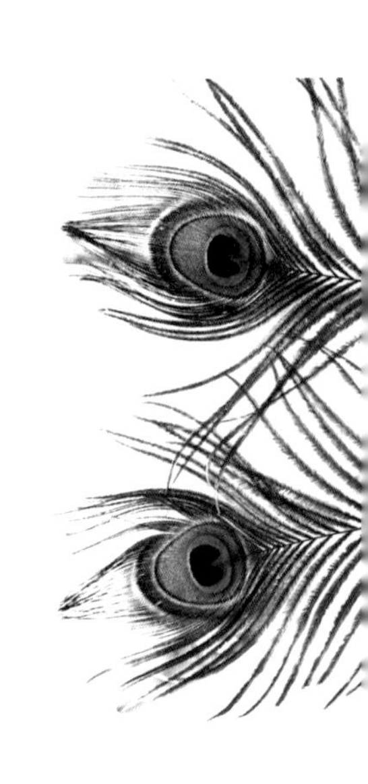

6.1 中国野生绿孔雀研究的历史

6.1.1 历史时期中国野生绿孔雀分布变迁

历史上关于中国野生绿孔雀的研究很少，据可查的记录最早是在1980年文焕然等的考证，发现绿孔雀在历史上曾遍布于湖南、湖北、四川、广东、广西和云南等省（自治区）。到21世纪初，绿孔雀在其他省区和云南东北部已绝迹，中国的分布区缩小到云南省的西部、中部和南部。

6.1.2 种群数量

据文贤继等（1995）于1991—1993年的信函调查与实地调查结果，绿孔雀种群数量在20世纪60年代以前最多，60年代以后由于栖息地的不断消失及滥捕滥猎，导致种群数量急剧下降，估计云南全省的绿孔雀种群数量在800~1100只，这是关于云南省绿孔雀种群数量较早、较全面的报道。1994—1995年对云南省西双版纳地区的绿孔雀开展调查，结果表明，西双版纳地区绿孔雀种群数量为19 ~25只（罗爱东和董永华，1998）。1999年对楚雄彝族自治州绿孔雀种群调查结果为280只（徐晖，1995）。在此后十多年间，尚未见到关于绿孔雀种群数量的研究报道。

文云燕等（2016）将红外相机法与标图法相结合，对楚雄双柏恐龙河州级自然保护区内的绿孔雀开展调查，结果显示，恐龙河保护区有56只绿孔雀。

2014年4月至2017年6月，孔德军和杨晓君采用问卷、访问和样线法对中国的绿孔雀资源进行调查，同时运用标图、鸣声、红外相机等方法进行补充，结果显示，中国绿孔雀种群数量已少于500只（孔德军和杨晓君，2017）。

滑荣等（2018）于2015—2017年通过信函调查与野外实地调查，估计我国绿孔雀种群数量为235~280只（表6-1）。在过去20年间，中国绿孔雀种群数量急剧减少，很多地区的绿孔雀都已消失，即使在有“孔雀之乡”称号的云南省西双版纳，也难以见到绿孔雀踪迹。

表6-1 历年报道的中国野生绿孔雀种群数量

年份	调查区域	绿孔雀种群数量（只）	资料来源
1995	云南省	800~1100	文贤继等，1995
1998	云南省西双版纳州	19~25	罗爱东和董永华，1998
1999	云南省楚雄州	约280	徐晖，1995
2016	云南省楚雄州恐龙河州级自然保护区	56	文云燕等，2016
2018	云南省	小于500	孔德军和杨晓君，2017
2018	云南省	235~280	滑荣等，2018

6.1.3 历史分布变化

在我国两广（广东、广西）、两湖（湖南、湖北）、四川和云南，历史上曾有绿孔雀广泛分布，20世纪初，云南东北部和云南以外其他省（自治区）的绿孔雀均已灭绝，分布区缩减至云南南部、中部和西部。

王紫江1983年的研究结果显示，绿孔雀在国内分布于云南省的泸水市、腾冲市、盈江县、孟定镇、西双版纳州和新平县等地；随后1990年报道，又在楚雄彝族自治州禄丰、双柏、南华、姚安4个县，以及楚雄市发现绿孔雀（王紫江，1990）。文贤继等（1995）1991—1993年对云南省绿孔雀调查结果表明，34个县（市）有绿孔雀分布，维西、德钦2个县尚待确认，有5个县过去有分布，现已灭绝或绝迹（表6-2）。杨晓君等（1997）1995—1996年对云南省东南部和西北部绿孔雀调查研究显示，云南东南部目前有绿孔雀分布的只有3个县：建水、石屏和弥勒；原来记录有绿孔雀分布，现在已经绝迹的有6个县（市）：文山、蒙自、金平、绿春、河口、开远。

罗爱东等（1998）1994—1995年对西双版纳州绿孔雀调查结果显示，绿孔雀仅分布于景洪市、勐海县和勐腊县的部分地区。

Han等2007年对云南省绿孔雀调查结果显示，云南省历史上有绿孔雀分布的42 个地区，盈江、腾冲、六库、蒙自和河口5个县（市）的绿孔雀在20世纪80年代末已经绝迹，勐腊、金平、绿春、建水4个县（市）在20世纪90年代已经绝迹，现绿孔雀仅分布于31个县（市）。

Kong等2014—2017年对云南省绿孔雀进行全面调查，在过去30年间，云南省绿孔雀分布区由11个州（市）34个县127个乡（镇），急剧缩减为8个州（市）22个县33 个乡（镇），西藏自治区的察隅县和墨脱县已没有绿孔雀分布。另外，滇中地区的元江县、峨山县是新发现的绿孔雀分布区（表6-2）。

表6-2　历次调查记录的中国绿孔雀分布地信息

时间	绿孔雀分布情况	数据来源
20世纪以前	湖南、湖北、四川、广东、广西、云南	文焕然和何业恒，1980
1983年、1990年	云南省：泸水、腾冲、盈江、孟定、新平、禄丰、双柏、南华、姚安、楚雄以及西双版纳等地有分布	王紫江，1983，1990
1995年	云南省：瑞丽、陇川、潞西（今芒市）、昌宁、龙陵、永德、镇康、耿马、沧源、双江、云县、临沧（今临翔区）、凤庆、新平、普洱（今宁洱县）、墨江、景东、景谷、镇沅、思茅（今思茅区）、楚雄（今楚雄市）、双柏、南华、永仁、姚安、禄丰、景洪（今景洪市）、勐海、勐腊、巍山、绿春、金平、石屏、弥勒34个县（市、区）有分布；盈江、泸水、腾冲（今腾冲市）、蒙自、河口5个县（市）已绝迹	文贤继等，1995
1997年	云南省东南部建水、石屏、弥勒3个县（市）有分布；蒙自、金平、绿春、河口、开远、文山等6个县（市）已经绝迹	杨晓君等，1997
2007年	云南省巍山、永仁、景洪、瑞丽、陇川、潞西、龙陵、昌宁、凤庆、云县、永德、镇康、耿马、沧源、双江、临沧、景东、景谷、镇沅、普洱（今宁洱）、思茅、勐海、墨江、石屏、弥勒、新平、双柏、楚雄、禄丰、南华、姚安等31个县（市、区）有分布	Han et al.,2007
2018年	云南省瑞丽、陇川、龙陵、昌宁、永德、镇康、耿马、澜沧、景谷、景东、宁洱、墨江、景洪、元江、新平、峨山、楚雄、南华、禄丰、双柏、石屏、建水22个县（市）有分布	Kong et al., 2018

王方等2017年1—6月，利用红外相机对云南省新平县野生动物进行连续监测。经统计，33台红外相机累计工作3836个工作日，捕获独立有效照片和视频1853组，其中鸟类1473组，兽类380组。经鉴定，拍摄到的鸟类有13种，隶属5目8科13属；兽类有8种，隶属5目8科8属。其中，国家Ⅰ级保护动物有2种，分别为绿孔雀(*Pavo muticus*)和黑颈长尾雉(*Syrmaticus humiae*)；国家Ⅱ级保护动物有5种，分别为原鸡(*Gallus gallus*)、白鹇(*Lophura nycthemera*)、白腹锦鸡（*Chrysolophus amherstiae*）、猕猴(*Macaca mulatta*)、鬣羚(*Capricornis milneedwardsii*)。IUCN红色名录濒危种有绿孔雀。采用G-F指数计算鸟兽多样性指数，定量分析红外相机拍摄的鸟兽科间、属间的多样性，鸟类的G-F 指数高于兽类，这表明该地区的鸟类多样性比兽类多样性高。鸟类相对丰富度较高的3个种是绿孔雀、原鸡和白鹇； 兽类相对丰富度较高的3类是鼠类、赤腹松鼠 (*Callosciurus erythraeus*) 和豹猫 (*Prionailurus bengalensis*)（王方等，2018a）。绿孔雀的种间关系主要包括捕食与被捕食、竞争、寄生、错配求偶等。

据王方等（2018）关于绿孔雀伴生鸟兽多样性及联结性关系研究，与绿孔雀伴生的鸟兽约20余种，包括原鸡、白鹇、黑颈长尾雉、白腹锦鸡、蓝喉拟啄木鸟、灰林鵖、绿翅金鸠、山斑鸠、紫啸鸫、黑领噪鹛、红嘴蓝鹊、黑卷尾等，也包括兽类赤麂、野猪、猕猴、豹猫、鬣羚、赤腹松鼠、云南兔等。

角媛梅等（2017）以拟建设的楚雄哀牢山国家公园碍嘉片区（即双柏县碍嘉镇）为对象，依据世界自然保护联盟(IUCN) 濒危物种绿孔雀的潜在生境进行功能分区。结果表明：① 楚雄哀牢山国家公园碍嘉片区的建设对哀牢山中山湿性常绿阔叶林和河谷季雨林等生态系统的完整性及其旗舰物种尤其是濒危物种绿孔雀的保护具有重要意义，可实现保护区和保护对象的空间整合；②依据海拔、坡度、坡向和植被类型提取的绿孔雀潜在生境和不适宜生境面积分别占总面积的 20.32%和79.68%，因潜在生境斑块面积小且破碎，所以将间距小于500 m 的潜在生境斑块进行连接，根据连接线的密度提取出潜在生境核心区；③在保护生态系统完整性和濒危物种适宜生境、整合原有保护区、兼顾社区发展和游憩利用的划分原则下，将碍嘉片区划分为核心保护区、生态保育区、传统利用区、游憩展示区与人类活动区，各区分别占总面积的66.90%、16.80%、7.01%、0.24%和9.05%，各功能区的保护利用要求不同。此外，还有关于绿孔雀栖息地、食性、饲养与繁殖、生理和遗传学、疾病及预防、受威胁因素和保护管理等方面的研究。在国外，印度尼西亚的爪哇岛（Jarwadi et al.，2011）和越南（Briekle，2002）等地开展了绿孔雀栖息地的研究，国内也较早地对景东县春季绿孔雀栖息地开展观察研究（杨晓君等，2000）；对绿孔雀的食性研究表明，绿孔雀是杂食性动物，以栎树、火把果、白株、豆科、菊科的嫩叶和花、草籽、稻谷等为食，喜欢吃棠梨、黄泡，也捕食蝗虫、金龟子科昆虫、蟋蟀、蚱蜢、小蛾、蛙类和蜥蜴等（匡邦郁，1963；文焕然和何业恒，1980；王紫江，1983；孔德军和杨晓君，2017）；饲养与繁殖主要是对绿孔雀饲养过程中饲料的成分、孵化方法及孵化过程中的温湿度、水分等进行研究（张春丽，1995；吴君，2004；李世强等，2006；王凤华等，2007；张仲安，2008；张丽霞等，2015）；生理和遗传方面主要是关于绿孔雀卵壳成分（王玉龙等，2000）、消化系统组织学观察（何平和陆宇燕，2002；李健等，2004）、血液生理指标（祁为伟等，1998；张云美等，2003；周庆萍等，2011）和绿孔雀分类地位（常弘等，2002；Ke et al.，2004；朱世杰等，2004；包文斌等，2006；欧阳依娜等，2009；段玉宝等，2018）的研究；疾病及预防主要关于绿孔雀组织滴虫病（王秀梅等，2001；闫港和侯俊丽，2002；史晓涛等，2008；周薇等，2015）、球虫病（张立春和李沐森，2008；李沐森和崔贞爱，2009；陈静，2011）、细菌感染（张建虎和邓妮娟，2002；赵恒章和李军民，2004；胡品昌等，2012；王武，2012）、新城疫疾病（吴长新等，1999）、肠道寄生虫病（胡艳，胡辉，2002；Foronda et al.，2004；刘云龙等，2011；黄超等，2015，2017）等疾病的发病特征及治疗措施；近年来，对绿孔雀受威胁因素及保护管理的研究综述较多，对绿孔雀的保护级别、濒危现状、受威胁因素和已开展的保护管理措施等方面进行报道（文贤继等，1995；杨晓君等，1997；谢以昌，2016；孔德军和杨晓君，2017；付昌健等，2019；顾伯健和陈雨茜，2019；李斌强等，2018）。

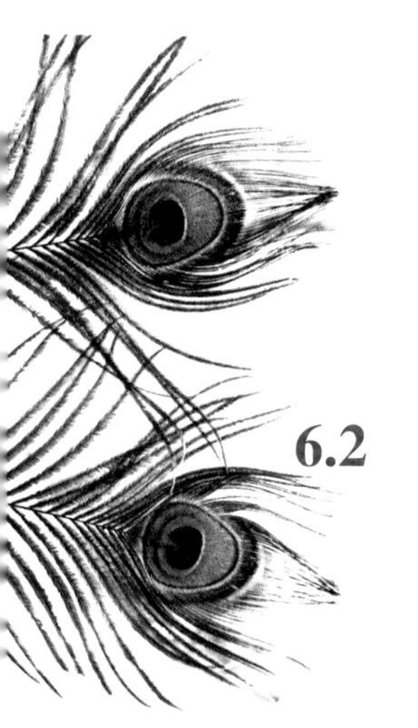

6.2 中国野生绿孔雀研究的现状

刘钊等在2007年3—4月和10—11月在云南元江上游石羊江河谷绿孔雀的分布区内，调查了绿孔雀的觅食生境，测定了21个生态因子（图6-3）。对绿孔雀栖息地的选择，不同季节觅食地选择，人为干扰对绿孔雀觅食地选择的影响等做了详细的调查研究。

▲ 图6-3 元江上游石羊江河谷绿孔雀分布区（王方/摄）

文云燕等（2015）在双柏县恐龙河州级自然保护区绿孔雀集中分布区域，利用标图法结合红外触发自动相机开展绿孔雀调查及监测，获得绿孔雀大量的野外图片及相关视频，初步确定了绿孔雀求偶、交配、孵蛋时间，并对保护区绿孔雀保护中存在的问题进行了分析（图6–4）。

▲ 图6–4　恐龙河流域绿孔雀分布区（王方/摄）

李旭等在（2016）以云南省楚雄州恐龙河自然保护区磨家湾绿孔雀为对象，分析了各种环境特征之间的差异，及其对绿孔雀栖息地选择的综合影响，结果表明，食物和隐蔽条件是影响绿孔雀觅食地选择的要素（图6–5）。

▲ 图6-5　云南省楚雄州恐龙河自然保护区（王方/摄）

孔德军等（2017）概述绿孔雀的基本生物学资料，包括形态特征、亚种分化、栖息地和繁殖特征、食性、种群数量与分布，同时指出我国绿孔雀保护所面临的问题（图6-6）。

▲ 图6-6　森林中觅食的野生绿孔雀（云南大学绿孔雀研究团队红外相机/摄）

单鹏飞等（2018）首次发现野生绿孔雀巢，并对绿孔雀巢址生境进行描述，同时对绿孔雀卵的重量及大小进行测量，与郑光美、杨岚等先前的描述一致。

王方等（2018）通过红外相机对云南省新平县野生绿孔雀进行监测，结果表明新平县有野生绿孔雀分布，并初步确定了分布范围及栖息地范围（图6-7～图6-9）。

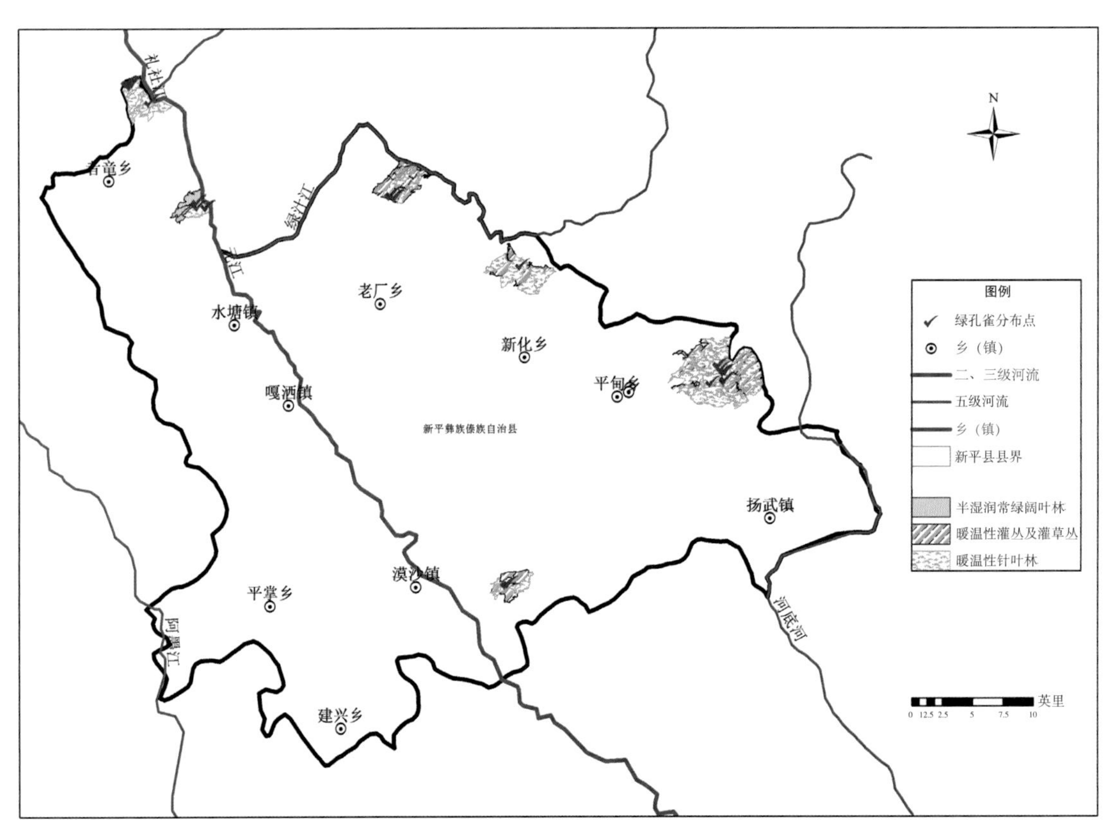

▲ 图6-7　云南大学绿孔雀研究团队完成的新平县绿孔雀分布现状图（陈明勇/制）

本图境界画法不作划界依据

▲ 图6-8 云南大学绿孔雀研究团队在样线上发现并测量绿孔雀足迹（陈明勇/摄）

▲ 图6-9 云南大学绿孔雀研究团队在野外布设红外相机（陈明勇/摄）

王方等（2018）通过长达一年的红外相机持续监测，获取了数万张野生绿孔雀的图片资料，并统计得出新平县野生绿孔雀种群数量为127~151只（图6-10）。

▲ 图6-10　用于监测绿孔雀的红外相机（王方/摄）

王方等（2018）对野生绿孔雀的伴生鸟兽多样性进行分析，并引入2×2列联表分析绿孔雀与伴生鸟兽的种间关系（图6-11）。

▲ 图6-11　红外相机拍摄到的绿孔雀食物竞争者——原鸡（云南大学绿孔雀研究团队红外相机/摄）

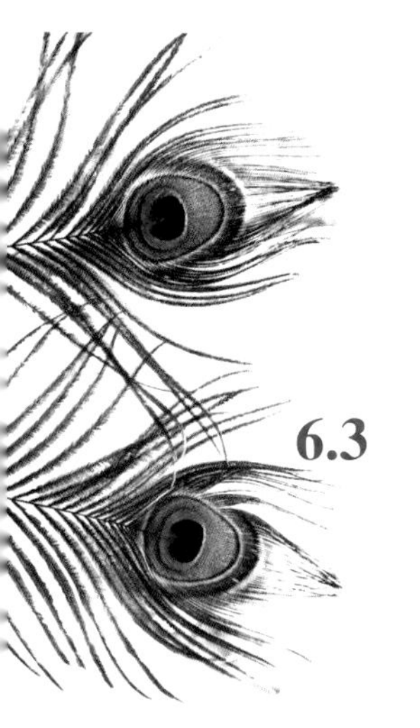

6.3　中国野生绿孔雀面临的主要问题

杨晓君等（2017）认为，中国野生绿孔雀主要的致濒危因素包括：致死（中毒、盗猎）、栖息地丧失（毁林、开矿、建水电站、修公路、种经济林等）、干扰（村庄、放牧、采摘）和保护管理（多分布于保护区外），而绿孔雀目前的种群状况（种群数量小、种群隔离、每个群体变小）、栖息地退化（图6-12）、人为干扰日益增大更加重了其濒危程度。

▲ 图6-12　栖息地退化（王方/摄）

滑荣等（2018）认为，我国绿孔雀面临的主要威胁有：①偷猎猖獗；② 毁林开荒、修路、水电站建设等导致的栖息地破坏；③栖息地破碎化导致小种群隔离分布，近亲繁殖；④农药包衣种子、灭鼠药、家禽传染病和种群内传染病；⑤当地村民笼养蓝孔雀逃逸，导致野生绿孔雀基因污染。

6.3.1 栖息地丧失严重

一是分布范围急剧萎缩。绿孔雀分布区由于替代种植、工程建设、森林砍伐、采矿、河床取沙、外来物种入侵等原因，使绿孔雀栖息地急剧缩减，与20世纪90年代相比，绿孔雀分布范围由42个县下降到19个县。二是栖息地破碎化严重。云南省绿孔雀种群均呈片段化分布，在种群密度较大的元江中上游地区，由于公路、电站、耕地、村寨、经济林等建设与阻隔，绿孔雀同样呈片段化、小群体分散化的分布格局，难以进行有效基因交流；目前绿孔雀重要栖息地元江中上游已建成了大湾电站、小江河一级电站，戛洒江一级电站、小江河二级电站虽暂停建设，但如果继续建设，将淹没海拔680 m以下的河谷地区，该区域的河滩目前是绿孔雀的沙浴地、求偶场地等重要场所，蓄水后绿孔雀适宜栖息地将被淹没而散失，而且大湾电站与戛洒江一级电站首尾相连，江水（库区）水面变宽，对东西两岸的绿孔雀种群造成隔离，严重影响种群之间的基因交流。三是栖息地适宜性下降。 由于农业生产方式的转变，农耕机、割草机等机械及农药使用，产生噪声和污染，对绿孔雀产生了消极影响;社区居民林下采摘、放牧等活动，干扰了绿孔雀的繁衍生息。

6.3.2 保护基础薄弱

一是管理机制不完善。绿孔雀主要沿红河、澜沧江和怒江流域分布，流域管理涉及水利、国土、林业、农业等部门的多头管理，管理权限分散，部门之间存在利益冲突，缺乏部门间的联动与协调管理机制，缺乏统一的流域管理理念，导致绿孔雀栖息地的破坏和退化。二是保护管理能力弱。在绿孔雀集中分布的双柏县和新平县仅建有恐龙河州级自然保护区，由于保护区级别比较低、资金投入缺乏、基础设施设备落后、人员编制不足、宣传教育滞后等原因，导致保护区管护管理能力弱。三是保护空缺较大。在云南省建立的各级自然保护区中，7个保护区还有绿孔雀分布，种群数量为170~183只，还有近2/3的绿孔雀种群数量在保护区外，保护空缺较大。

6.3.3 科研监测基础薄弱

一是基础性研究薄弱。国家、地方对绿孔雀的研究投入经费较少，绿孔雀栖息地利用和选择、繁殖、种群遗传结构等研究数据缺乏，难以根据绿孔雀的生态学、生物学特性提出针对性的保护管理措施。二是监测体系不完善。绿孔雀监测仅在恐龙河州级保护区的局部区域开展，全省未建立完整的监测体系，监测手段主要靠人工巡护，缺乏现代化监测设备和手段，监测能力难以满足实际需要。

6.3.4 宣传教育滞后

一是宣传教育开展较少。由于绿孔雀保护宣传教育工作开展较少，公众很难识别外来种蓝孔雀和绿孔雀，个别媒体、商家在利用孔雀形象进行商品宣传时更是将二者混淆，导致公众对当前绿孔雀珍稀濒危程度、保护级别，以及保护所面临的风险和困境认识不足；绿孔雀分布区多位于边远贫困山区，社区保护意识淡薄。二是缺乏生态农业种植引导。随着农业生产方式的改变，农户播撒带有农药的“包衣种子”或浸泡过农药的作物种子、在田地里喷洒农药、投放鼠药等，均会导致绿孔雀误食而死亡。

6.3.5 资金投入不稳定

2009 年以来，林业部门仅投入资金 300 多万元，其中，2017 年国家林业局下达专项保护资金 165 万元，2017 年、2018 年云南省林业和草原局分别划拨元江上游绿孔雀种群和栖息地调查与评估经费 19.5 万元、云南省绿孔雀保护实施方案编制经费 10 万元。由于国家野生动物保护专项资金较少，云南省没有专项保护经费，绿孔雀栖息地管理、监测、公众教育等工作难以有效开展。

据分析，云南省玉溪市新平县野生绿孔雀正经历着持续的栖息地丧失和种群数量下降的威胁，拯救保护绿孔雀工作已经迫在眉睫。而造成中国野生绿孔雀分布变迁的主要原因有放牧、大型工程项目建设、偷猎、栖息地退化、人为干扰等外部因素和绿孔雀自身的种质因素等（图6-13 ~ 图6-15）。

▲ 图6-13　绿孔雀分布区周边大量开垦的种植园（王方/摄）

如果我们对每一类威胁因子进行分析，分别从范围 (scope)、严重性(seventy)和不可逆转性(irreversibility)等3个方面进行赋值，分析得出目标-威胁级别最高的是栖息地破坏；其次是放牧、采沙、开矿、狩猎等；再次为采伐、公路、耕作、林下采集、生火等，这些威胁因子都是中国野生绿孔雀濒危的原因所在。

▲ 图6-14　绿孔雀分布区开发后的种植园（王方/摄）

▲ 图6-15 绿孔雀分布区工程建设现场（王方/摄）

随着经济的发展，水电站、发电厂等大型工程项目的建设也严重破坏了绿孔雀的栖息地，还造成了绿孔雀种群间的地理隔离，加之种植、非法偷猎、伐木等人为干扰，绿孔雀的种群数量以及栖息地的分布范围仍在减小（图6-16）。

▲ 图6-16 绿孔雀分布区大量采代的木材（王方/摄）

每年大量的木材砍伐不仅造成绿孔雀栖息地的减少，同时也对绿孔雀的生存造成较大干扰（图6–17）。

▲ 图6–17　在绿孔雀栖息地人工割松脂的痕迹（王方/摄）

红河流域大量的采沙船持续开展着采沙、洗沙工作，使江水变得浑浊，发出的噪声也严重影响着周边绿孔雀的生活（图6–18）。

▲ 图6–18　在绿孔雀栖息地河流有许多采沙船（王方/摄）

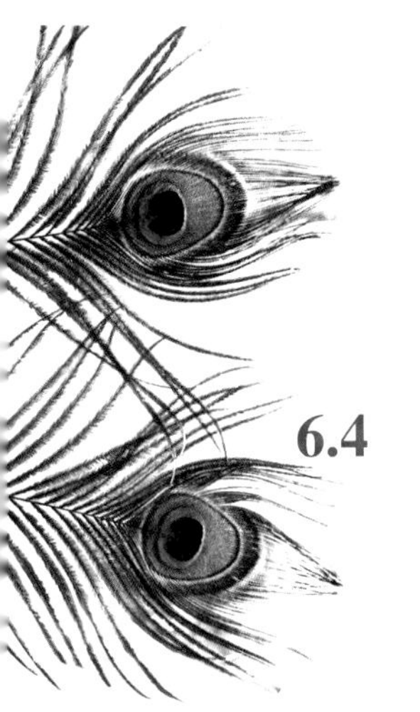

6.4 中国野生绿孔雀保护现状

中国野生绿孔雀的保护在分布区各级政府和部门之间尚未形成一致意见，且各个保护区未配备专业人员，管理人员未能对绿孔雀有全面科学的认识，人员缺口大。各个管理区保护局监测设备不全或老旧，难以全面快速地对绿孔雀进行研究活动。目前，中国野生绿孔雀的保护工作得到了社会各界的广泛关注。为了更好地保护中国野生绿孔雀，云南省将其列为极小种群拯救保护物种，对其进行重点保护和拯救。中国科学院昆明动物研究所开展了全省范围内野生绿孔雀现状分布调查工作。阿拉善组织和云南省林业和草原局在云南省玉溪市新平县开设绿孔雀保护小区，云南大学、西南林业大学等高校也开始集中鸟类学研究科研力量对野生绿孔雀开展调研保护工作（图6-19 ~ 图6-21）。

▲ 图6-19　开展绿孔雀野外调查（王方/摄）

▲ 图6-20　绿孔雀保护宣传牌（王方/摄）

▲ 图6-21 绿孔雀保护专题论证会（曹顺/摄）

7

孔雀文化

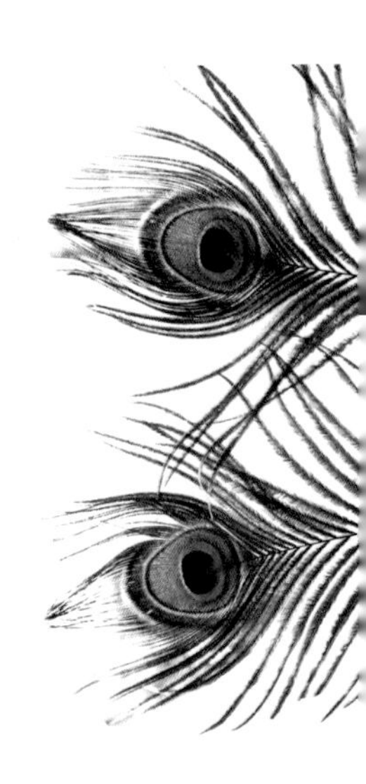

7.1 文化价值

绿孔雀从古至今就和人类有着深刻的历史渊源，与我国少数民族之间的联系更是盘根错节，且在文学、神话、宗教、民族艺术等方面颇负盛名（图7-1）。

▲ 图7-1 《孔雀红梅图》（宋·佚名）

人们熟知的汉乐府《孔雀东南飞》中言“汉末建安中，庐江府小吏焦仲卿妻刘氏为仲卿母所遣，自誓不嫁，其家逼之，乃投水而死，仲卿闻之，亦自缢于庭树”故事，开篇说的“孔雀东南飞，五里一徘徊”，又有“两家求合葬，合葬华山傍”句。

在中国的神话及宗教传说中，孔雀又是百鸟之王凤凰得交合之气后孕育而生，与大鹏金翅鸟为一母同胞，后被如来佛祖敕封为“孔雀大明王菩萨”，故又称“孔雀大明王”（王研博，2018）。

在云南，很多少数民族也极其喜欢孔雀，如滇中的彝族、滇南的傣族等，都把孔雀作为重要的图腾来崇拜。在傣族人民的心目中，孔雀是善良且聪明的，是最爱自由与和平的鸟，是吉祥、幸福的象征。

傣族文化有着1300多年的历史，亚热带湿热的气候让傣族人民过着安静的农耕生活。傣族人民大多信奉小乘佛教，追求智慧和空灵的境界，与孔雀的灵动与娴静很是相像。

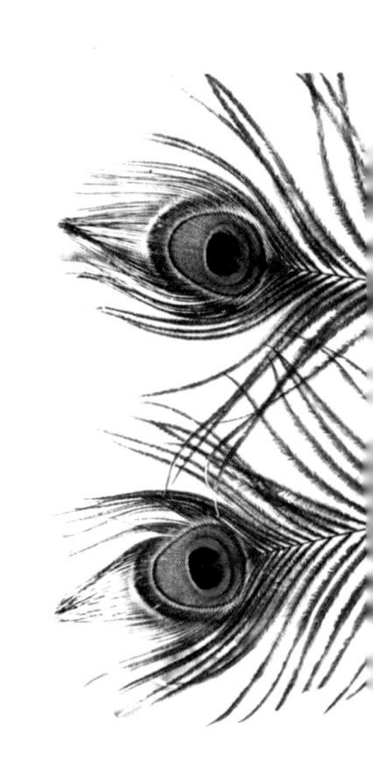

7.2 孔雀与凤凰

凤凰有多种原型，如锦鸡、孔雀、鹰鹫、鹄、玄鸟等。作为凤凰的原型，孔雀是傣族崇拜的吉祥鸟。傣族地区孔雀较多，据史书记载，“孔雀巢人家树上”。可见自古孔雀就与傣家人结下了不解之缘，西双版纳傣族自治州和德宏傣族景颇族自治州都被誉为“孔雀之乡”。傣族群众常把孔雀作为自己民族精神的象征，并以跳孔雀舞来表达自己的愿望和理想，歌颂美好的生活。孔雀舞是傣族古老的民间舞，也是傣族人民最喜欢的舞蹈，流行于整个傣族地区，其中以瑞丽、耿马、孟定、勐腊等地的孔雀舞最为精彩。过去表演者多戴面具，他们模仿孔雀漫步森林、饮泉戏水、追逐游戏、抖翅展翅、亮翅开屏、高低旋转等动作。

孔雀在傣族文学中是最具有灵性的动物，如在民间故事《两个王子的故事》中，王子两弟兄因用箭射死了母雁而被公雁告发，国王定两位王子死罪，在王后的请求下，执行人员放了两位王子。两位王子流浪异国，射死了那一国家国王的孔雀，大王子吃了孔雀的头浑身闪闪发光，成为国王，小王子吃了孔雀的肝，说话时金银财宝从口中喷出……现今傣族以孔雀为吉祥物，跳孔雀舞，以孔雀为织锦、刺绣、剪纸、绘画的重要主题，以孔雀图样为房屋的装饰（图7-2、图7-3）或妇女的饰品等。这一切，与早期的傣族祖先——古越人的鸟图腾崇拜密切相关。傣族还以孔雀为文身图案，这显然是希望得到图腾物凤凰的原型——孔雀的保佑，幸福平安，吉祥如意。傣族英雄史诗《厘俸》中的英雄“桑洛”的军旗上绣有孔雀，西双版纳宣慰使的仪仗旗上有凤凰图案，以孔雀或凤凰为集团政权的徽征，这与傣族早期的鸟图腾崇拜不无关系（刀承华，2009）。

▲ 图7-2 傣寨缅寺建筑上的孔雀雕塑（汤永晶/摄）

▲ 图7-3 傣寨缅寺中佛祖身边的绿孔雀、白孔雀和白象雕塑（汤永晶/摄）

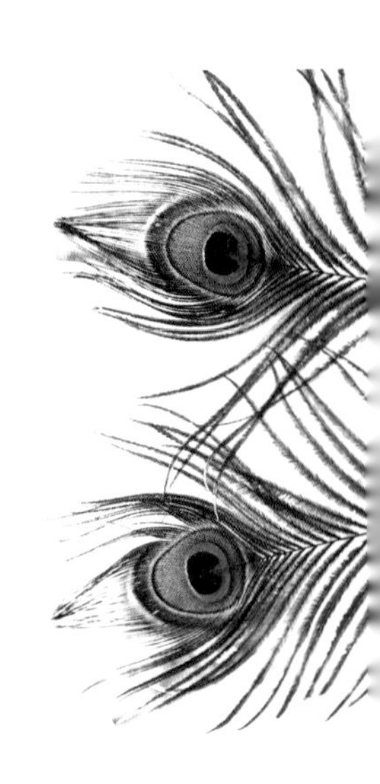

7.3 孔雀舞

孔雀舞在傣族地区有着悠久的传承基础，傣乡缅寺的古老壁画和雕刻中，都有表现栩栩如生的人面鸟身的孔雀形象，这与傣乡现有的头戴尖塔和假面具、身着孔雀服的孔雀舞十分相似，可见孔雀舞的历史源远流长。孔雀舞，傣语称“嘎洛涌”“烦洛涌”或“嘎楠洛”，孔雀舞便成为了傣族人民最为喜闻乐见的舞蹈。傣乡素有“孔雀之乡”的美称，过去每当晨曦微明或夕阳斜照时，常见姿态旖旎的孔雀翩翩起舞。明《南诏野史》曾这样记载：“婚取长幼跳舞，吹芦笙为孔雀舞。”

人们以孔雀翎献佛，跳孔雀舞求吉祥，孔雀舞自然而然地成为傣族舞文化与审美的灵魂（图7-4 ~ 图7-7）。

▲ 图7-4　傣族孔雀舞蹈（图片引自长谷川清）

▲ 图7-5　傣族白孔雀舞蹈（图片引自长谷川清）

▲ 图7-6 西双版纳傣族泼水节期间的孔雀装游行（陈明勇/摄）

▲ 图7-7 柬埔寨当地民族的孔雀服饰（陈明勇/摄）

7.4 新疆“孔雀”名物考与“孔雀河”名的由来

王守春（2015）对新疆孔雀名物考与“孔雀河”名的由来进行了专题研究，经考证后认为，《魏书·西域传》记载龟兹国有“孔雀”，实际上它并非孔雀，而是被当地居民称为“Kum-tuche”（汉语音译为“沙图提”）的鸟，“Kum” 在维吾尔语中为“沙”之意，为意译，“图提”则为“tuche”的音译。“沙图提”实际上就是“Kum-tuche”意译和音译相结合的混合译名，其确切的汉语意译，应是“沙雀”或“沙地之雀”，表明这种鸟主要应是生存在沙地或荒漠环境中。在现代鸟类分类系统中属于雀形目鸦（原文误为“鸭”）科中的几种鸟，可能包括了黑尾地鸦（原文误为“黑尾地鸭”）*Podoces hendersoni* Hume、褐背地鸦（原文为误为“褐背地鸭”）*Podoces humilis* Hume、白尾地鸦（原文误为“白尾地鸭”）*Podoces biddulphi* 和红嘴山鸦（原文误为“红嘴山鸭”，*Pyrrhocorax pyrrhocorax*）。因为这种鸟的习性和生态特点与《魏书》所记“孔雀”也很接近；而《新疆游记》一书中所描写的“孔雀”很可能就是红嘴山鸦（原文误为“鸭”），因上述三种地鸦都主要是在山丘之间或戈壁荒漠的环境中生存，只有红嘴山鸦是在高山环境中生存。历史上，被塔里木盆地人民称为“Kum-tuche”的鸟与当地的民俗有密切关系。除了《魏书》记载龟兹国人食用“Kum-tuche”，还因为黑尾地鸦的羽毛具有美丽的紫蓝色或紫黑色的金属光泽，故从前维吾尔族妇女将其插在帽子上作为装饰。历史文献记载的塔里木盆地的“孔雀”也不应是绿孔雀，而是新疆当地的“Kum-tuche”鸟（即黑尾地鸦）。

孔雀河（维吾尔语音译为“昆其达利雅”）是新疆的一条重要河流，发源于博斯腾湖，它的水源是开都河水入湖自然调节后，从大湖的西南角流出，经过苇湖区汇集成孔雀河。“孔雀河”一名最初是指库尔勒以下的河段。孔雀河在清代中期的名称“开都河”或“海都河”一名，是按照河道的特点来取名的。从新疆地名取名的惯例来看，许多地名或水体名或是根据自然特点来取，如“开都河”是根据孔雀河在库尔勒西面形成的弯曲命名，或取自动物或植物，如布它海子因终年有水鸟栖息而得名，“孔雀河”可能也是“Kum-tuche”鸟（即黑尾地鸦）来命名。

▲ 图7-8 大型孔雀雕塑（陈明勇/摄）

8 展望

8.1 加强绿孔雀基础生态学研究

翔实的绿孔雀生物学、生态学、保护生物学研究成果是制定科学保护政策和策略、实施保护管理措施的前提和基础。鉴于目前我国野生绿孔雀基础研究还较薄弱，建议组织相关科研单位、高等院校科研人员，定期开展全域绿孔雀种群分布及数量调查，种群活动规律和种群扩散研究、食性和行为学研究、栖息地选择研究、繁殖生态学研究、潜在栖息地预测、栖息地恢复关键技术研究、人为干扰对绿孔雀种群的影响、绿孔雀保护生物学、绿孔雀种群遗传多样性、绿孔雀疫源疫病等方面的科学研究，夯实绿孔雀生态学基础资料，为制定相关管理政策和保护措施提供科学的理论依据。

8.2 加强绿孔雀种群保护管理

在绿孔雀种群集中分布区扩大现有自然保护区面积，强化自然保护区机构和能力建设，制定自然保护区保护管理计划及绿孔雀专门保护计划，加大巡护监测力度和执法力度。

严禁猎杀、捕捉和伤害野生绿孔雀，加大绿孔雀分布区森林公安、林业和草原、自然保护区建设和执法力度，定期、不定期开展专项打击非法盗猎绿孔雀和其他野生动物活动。通过保护、监测、人工增殖、科普教育等一系列措施，确保绿孔雀种群数量稳步增长；通过绿孔雀栖息地保护与修复，实现绿孔雀栖息地得到有效保护；通过积极鼓励社会力量广泛参与保护，加大公众正面宣传力度，有效提高社会参与度和公众教育成效。

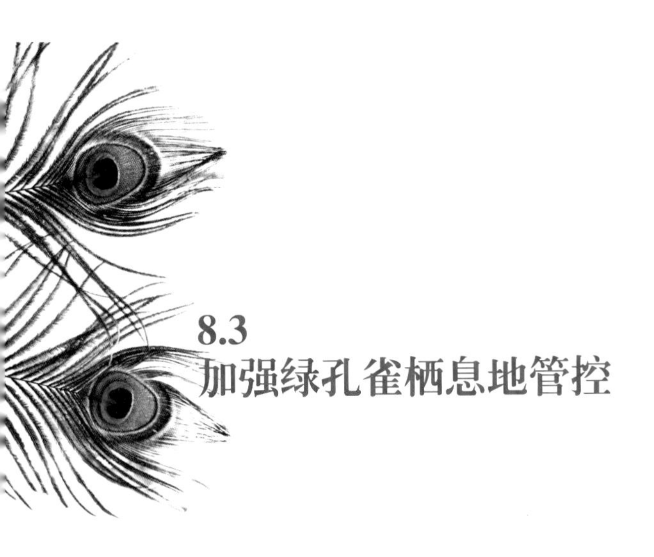

8.3 加强绿孔雀栖息地管控

在现有绿孔雀种群分布区域间划建绿孔雀保护廊道，对现有的分布区栖息地质量进行调查和评估，根据各区域特征制定栖息地保护、恢复方案，并将栖息地保护纳入政府工作计划和财政预算。

划定并建设专门的绿孔雀国家公园、自然保护区或保护小区，加大绿孔雀原生境的保护；对涉及绿孔雀栖息地的大型基地建设工作要制定影响评价标准，严格审批程序；禁止在潜在栖息地和生态廊道范围内采石、采沙、采薪、开荒等人为活动；禁止在潜在栖息地和生态廊道周边进行采矿、采石的爆破活动；禁止在潜在栖息地和生态廊道范围内宿营、穿越等人为活动；落实管护、巡护人员，严防人为破坏和森林火灾；实施人工和红外相机监测。

8.4 加强野生绿孔雀保护宣传教育

构建绿孔雀及其栖息地保护宣传体系，如建设自然保护博物馆、科技馆、文化馆，以展览、自然教育、科普宣教等活动形式开展绿孔雀及其栖息地保护宣传。组织专家编写保护绿孔雀的宣传手册及其他宣传材料，组织开展大学、中学、小学及幼儿园学生的宣传教育，提高保护意识。利用微信、微博平台，通过电影、录像、宣传片、纪录片等形式开展公众保护意识宣传，通过发放宣传资料、专题讲座、家访等形式走进学校、社区宣传绿孔雀保护、自然保护相关法律、法规知识，增强社会各界、公众，尤其是绿孔雀分布区居民热爱绿孔雀、保护绿孔雀的积极性，提高全社会的保护意识。

8.5 积极开展绿孔雀人工繁育与野化放归

在曾经有野生绿孔雀分布的自然保护区和栖息地，开展绿孔雀人工繁育与野化放归对于重建当地绿孔雀种群具有重要意义。规划和建设绿孔雀繁育和野化基地，由对绿孔雀及大型雉类有深入研究的科研院所、高等院校作为科技支撑，林业和草原、自然保护区科技人员参与，通过科学研究与试验，绿孔雀人工繁育技术有所突破，实现绿孔雀种群数量的人工增长。

参考文献

[1] 艾怀森. 高黎贡山地区雉类多样性及其保护[J]. 动物学研究, 2006, 27 (4):427-432.

[2] 白娜. 从史料探究中国历史时期孔雀的地理分布[J]. 文山学院学报, 2015, 28(4):56-58.

[3] 包文斌, 陈国宏, 束靖婷, 等. 孔雀微卫星引物筛选及其遗传多样性分析[J].遗传, 2006, 28(10):1242-1246.

[4] 陈静. 绿孔雀球虫病的诊断与治疗措施[J]. 养殖技术顾问, 2011(11):217.

[5] 常弘, 柯亚永, 苏应娟, 等. 野生与笼养绿孔雀种群的随机扩增多态DNA研究[J]. 遗传, 2002, 24(3):271-274.

[6] 丛培昊, 郑光美. 红腹角雉(*Tragopan temminckii*)的孵卵和育雏行为研究[J]. 北京师范大学学报(自然科学版), 2008, 44(4):405-410.

[7] 崔鹏, 康明江, 邓文洪. 繁殖季节同域分布的红腹角雉和血雉的觅食生境选择[J]. 生物多样性, 2008, 16(2):143-149.

[8] 党心言, 马国强, 唐杨春, 等. 动物园人工饲养蓝孔雀营养状况分析[J]. 西南林业大学学报, 2014, 34(1):102-105.

[9] 段玉宝, 李媛, 陈熙, 等. 基于线粒体基因分析蓝孔雀与绿孔雀的遗传差异[J]. 基因组学与应用生物学, 2020, 39(2):547-552.

[10] 冯锋, 毕延台, 何相宝. 笼养绿孔雀繁殖期行为观察[J]. 野生动物学报, 2007, 28(4):16-17.

[11] 付昌健, 邱焕璐, 宇佳.中国绿孔雀濒危现状及其保护[J]. 野生动物学报, 2019, 40(1):233-239.

[12] 郭宝用, 王志胜, 兰道英. 南滚河保护区野生动物资源现状[J]. 野生动物, 1999, 20(4):46-47.

[13] 郭冬生, 张正旺. 中国鸟类生态大图鉴[M]. 重庆：重庆大学出版社, 2015.

[14] 王娇, 张崇良, 薛莹, 等. 海州湾及其邻近水域主要鱼类种间关联性[J]. 应用生态学报, 2020, 31(1):293-300.

[15] 何业恒. 中国珍稀鸟类的历史变迁[M]. 长沙：湖南科学技术出版社, 1994.

[16] 何平, 陆宇燕. 绿孔雀消化系统的组织学观察[J]. 陕西师范大学学报(自然科学版), 2002, 30(4):92-95.

[17] 黄超, 鲁永超, 杨娟, 等. 圆通山动物园绿孔雀肠道寄生虫种类调查[J].畜牧与饲料科学, 2015, 36(10):8-9.

[18] 黄超, 刘安荣, 杨娟, 等. 动物园绿孔雀肠道寄生虫流行特征调查[J]. 中国兽医杂志, 2017, 53(1):53-55.

[19] 滑荣, 崔多英, 刘佳, 等. 中国绿孔雀种群现状调查[J]. 野生动物学报, 2018, 39 (3): 681-684.

[20] 贾兰坡, 张振标. 河南淅川县下王岗遗址中的动物群[J]. 文物, 1977(6):41-49.

[21] 角媛梅, 刘歆, 李绒, 等. 基于绿孔雀潜在生境的楚雄哀牢山国家公园功能分区研究[J]. 旅游科学, 2017, 31(3):75-84.

[22] 匡邦郁. 云南南部的孔雀[J]. 生物学通报, 1963, 11(4):17-18.

[23] 孔德军, 杨晓君. 绿孔雀及其在中国的保护现状[J]. 生物学通报, 2017, 52(1):9-11, 64.

[24] 蓝勇. 中国历史地理[M]. 北京：高等教育出版社, 2002.

[25] 李斌强, 李鹏映, 杨家伟, 等. 运用红外相机调查云南巍山青华绿孔雀自然保护区的鸟兽多样性[J]. 生物多样性, 2018, 26 (12): 1343-1347.

[26] 李健, 王丽萍, 王文. 2月龄绿孔雀消化系统组织学观察[J]. 东北林业大学学报, 2004, 32(2):62-64.

[27] 李明富, 李晟, 王大军, 等. 四川唐家河自然保护区扭角羚冬春季日活动模式研究[J]. 四川动物, 2011, 30(6): 850-855.

[28] 李旭, 刘钊, 周伟, 等. 云南楚雄恐龙河保护区绿孔雀春季栖息地选择和空间分布[J]. 南京林业大学学报, 2016, 40(3):87-93.

[29] 李瑞年, 林海晏. 瑞丽至孟连高速公路建设项目沿线绿孔雀分布初探[J].中国环境科学学会科学技术年会论文集, 2018:493-497.

[30] 刘芳, 李琳娜, 何葆杰, 等. 绿孔雀的饲养与繁殖[J]. 特种经济动植物, 2012, 3:12-13.

[31] 刘小斌, 韦伟, 郑筱光, 等. 红腹锦鸡和红腹角雉活动节律——基于红外相机监测数据[J]. 动物学杂志, 2017, 52(2):194-202.

[32] 刘钊, 周伟, 张仁功, 等. 云南元江上游石羊江河谷绿孔雀不同季节觅食地选择[J]. 生物多样性, 2008, 16(6):539-546.

[33] 卢汰春. 中国珍稀濒危野生鸡类[M]. 福州：福建科学技术出版社, 1991.

[34] 罗爱东, 董永华. 西双版纳野生绿孔雀种群数量及分布现状调查[J]. 生态学杂志, 1998, 17(5):6-10.

[35] 马建章. 中国野生动物保护实用手册[M]. 北京：科学技术文献出版社, 2002.

[36] 欧阳依娜, 杨泽宇, 李大林, 等. 利用细胞色素b(Cyt b)基因分析绿孔雀和蓝孔雀的遗传分化[J]. 云南农业大学学报, 2009, 24(2):220-224.

[37] 宋志勇, 杨鸿培, 余东莉, 等. 西双版纳灰孔雀雉的种群数量现状研究[J]. 西部林业科学, 2018, 7(6):67-72.

[38] 单鹏飞, 吴飞. 绿孔雀 难得一见[J]. 森林与人类, 2017, (2):62-63.

[39] 唐松元, 李立, 段文武, 等. 蓝孔雀人工繁育技术[J]. 湖南林业科技, 2019, 46(3):75-79.

[40] 王研博. 历史时期中国孔雀分布变迁及其文化研究[D]. 西北农林科技大学, 2018.

[41] 文焕然, 何业恒. 中国古代的孔雀[J]. 化石, 1980 (3):8-9.

[42] 文焕然, 何业恒. 中国历史时期植物与动物变迁研究[M]. 重庆：重庆出版社, 1995.

[43] 文焕然, 何业恒. 中国历史时期孔雀的地理分布及其变迁[A]. 见：文焕然等著, 文榕生选编整理. 中国历史时期植物与动物变迁研究[M]. 重庆：重庆出版社, 2006.

[44] 文贤继, 杨晓君, 韩联宪, 等. 绿孔雀在中国的分布现状调查[J]. 生物多样性, 1995, 3(1):46-51.

[45] 文云燕, 谢以昌, 李学红. 恐龙河州级自然保护区绿孔雀监测探讨[J]. 林业调查规划, 2016, 41(4):69-71.

[46] 吴君. 绿孔雀饲养与繁殖状况的调查与分析[J]. 黑龙江动物繁殖, 2004, 12(2):46-47.

[47] 徐晖. 楚雄州绿孔雀的分布现状及保护措施[J]. 云南林业科技, 1995, 72(3):48-52.

[48] 徐晖. 云南楚雄绿孔雀的现状[J]. 野生动物, 1999, 20(3):12-13.

[49] 郑光美. 中国雉类[M]. 北京：高等教育出版社, 2015.

[50] 郑作新. 中国经济动物志・鸟类[M]. 北京：科学出版社, 1963.

[51] 朱志飞, 沈曼曼, 曹卉, 等. 蓝孔雀繁殖期行为观察和分析[J]. 上海畜牧兽医通信, 2011 (2):28-29.

[52] 朱世杰, 常弘, 张国萍, 等. 孔雀属孔雀线粒体细胞色素b基因全序列分析及其系统进化研究[J]. 中山大学学报, 2004, 43(6):45-47.

[53] 王守春. 新疆孔雀名物考与孔雀河名的由来[J]. 西域研究, 2015(2):22-29.

[54] 王子今. 龟兹“孔雀”考[J]. 南开学报（哲学社会科学版）, 2013(4):81-88.

[55] 王研博, 郭风平. 环境史视野下中国孔雀的分布与变迁及原因探讨[J]. 保山学院学报, 2018, 37(1):29-36.

[56] 王玉龙, 赵广英, 田秀华, 等. 3种孔雀卵壳成分分析及超微结构观察[J]. 黑龙江畜牧兽医, 2000, (4):36-37.

[57] 王紫江. 云南孔雀[J].云南林业, 1983(4):30.

[58] 王紫江. 云南楚雄发现孔雀[J]. 动物学研究, 1990, 11(1):54.

[59] 王方, 姚冲学, 刘宇, 等. 基于红外触发相机技术的新平县野生绿孔雀分布调查[J]. 林业调查规划, 2018, 43(6):10-14, 111.

[60] 王方, 蒋桂莲, 张志中, 等. 云南省新平县野生绿孔雀伴生鸟兽多样性及关联性分析[J]. 野生动物学报, 2018, 39(4):812 -819.

[61] 徐晖. 楚雄州绿孔雀的分布现状及保护措施[J]. 云南林业科技, 1995, 72(3):48-52.

[62] 鲜方海, 喻晓钢. 四川唐家河保护区内野生红腹锦鸡生物学特性及繁殖习性研究[J]. 四川动物, 2008, 27(6):1175-1178.

[63] 谢以昌. 恐龙河自然保护区绿孔雀保护思考[J]. 林业调查规划, 2016, 41(4):69-72.

[64] 谢以昌. 恐龙河自然保护区生物资源现状及保护对策[J]. 林业调查规划, 2009(1):10-12.

[65] 新平彝族傣族自治县地方志编纂委员会办公室. 新平年鉴[M]. 芒市：德宏民族出版社, 2017.

[66] 许龙, 张正旺, 丁长青. 样线法在鸟类数量调查中的运用[J]. 生态学杂志, 2003, 22(5):127-130.

[67] 杨岚. 云南鸟类志・上卷・非雀形目[M]. 昆明：云南科技出版社, 1995.

[68] 杨晓君, 孔德军, 吴飞, 等. 中国绿孔雀的种群现状与保护[C]. 第十三届全国野生动物生态与资源保护学术研讨会暨第六届中国西部动物学学术研讨会, 2017-10-27.

[69] 杨晓君, 杨岚. 笼养绿孔雀行为活动时间分配的初步观察[J]. 动物学报, 1996, 42(S1):106-111.

[70] 杨晓君, 文贤继, 杨岚. 云南东南部和西北部绿孔雀分布的调查[J]. 动物学研究, 1997, 18 (1):12, 18.

[71] 杨晓君, 文贤继, 杨岚, 等. 春季绿孔雀的栖息地及行为活动的初步观察[C]. 中国鸟类学研究第四届海峡两岸鸟类学术研讨会文集, 2000, 64-70.

[72] 杨忠兴, 王勇, 华朝朗, 等. 云南省绿孔雀保护存在问题及对策[J]. 福建林业科技, 2019, 46(4):112-119.

[73] 约翰・马敬能, 卡伦・菲利普斯, 何芬奇. 中国鸟类野外手册[M]. 长沙：湖南教育出版社, 2000.

[74] 严晓娟. 蓝孔雀肉用品质特性研究[D]. 甘肃农业大学, 2009.

[75] 余玉群, 张陕宁, 巩会生. 佛坪自然保护区雉类分布和密度的初步调查[J]. 野生动物, 1990, 57(5): 16-18.

[76] 袁景西, 张昌友, 谢文华, 等. 利用红外相机技术对九连山国家级自然保护区兽类和鸟类资源的初步调查[J]. 兽类学报, 2016, 36(3):367-372.

[77] 张春丽. 绿孔雀的饲养与繁殖[J]. 野生动物, 1995, (4):16-18.

[78] 张丽霞, 卫泽珍, 吴菲, 等. 绿孔雀的人工孵化和育雏[J]. 饲料博览, 2015, (7):41-44.

[79] 张云美, 吴登虎, 杨晓黎, 等. ILT对孔雀部分血液化学指标的影响观察[J]. 医学动物防制, 2003, 19(12):705-707.

[80] 曾昭漩. 试论珠江三角洲地区象、鳄、孔雀灭绝时期[J]. 华南师院学报（自然科学版）, 1980 (1):173-185.

[81] 赵玉泽, 王志臣, 徐基良, 等. 利用红外照相技术分析野生白冠长尾雉活动节律及时间分配[J]. 生态学报, 2013, 33(19):6021-6027.

[82] 郑光美. 世界鸟类分类与分布名录[M]. 北京：科学出版社, 2002.

[83] 郑光美. 中国鸟类分类与分布名录（第三版）[M]. 北京：科学出版社, 2017.

[84] 郑光美, 王岐山. 鸟类[A]. 见：汪松主编. 中国濒危动物红皮书[M]. 北京：科学出版社, 1998.

[85] 郑作新. 中国鸟类种和亚种分类名录大全[M]. 北京：科学出版社, 2000.

[86] 中华人民共和国林业部. 国家重点保护野生动物名录[M]. 北京：中国林业出版社, 1989.

[87] 周庆萍, 陈红, 周雪林, 等. 绿孔雀血液生化指标的测定[J]. 湖北农业科学, 2011, 50(15):3124-3126, 3130.

[88] 周晓禹, 王晓明, 姜振华. 贺兰山石鸡越冬期昼夜行为时间分配及活动节律[J]. 东北林业大学学报, 2008, 36(5): 44-46.

[89] 朱自强. 孔雀的饲养管理与繁殖[A]. 见：中华人民共和国濒危物种进出口管理办公室主编. 中国濒危经济野生动物驯养繁殖[M]. 哈尔滨：东北林业大学出版社, 1997.

[90] Briekle N W. Habitat use, predicted distribution and conservation of green peafowl *Pavo muticus* in Dak Lak Province, Vietnam[J]. *Biological Conservation*, 2002, 105:189-197.

[91] CITES. Convention on International Trade in Endangered Species of Wild Fauna and Flora[EB /OL]. https://www.cites.org/（2020-02-01）.

[92] Dejun Kong, Fei Wu, Pengfei Shan, et al. Status and distribution changes of the endangered Green Peafowl (*Pavo muticus*) in China over the past three decades (1990s-2017) [J]. *Avian Research*, 2018, 9(2):102-110.

[93] Dong F, Kuo HC, Chen GL, et al. Poputation genomic climat and anthropogenic evidence suggest the role of human forces in endengerment of green peafowl (*Povo muticus*)[J]. Proceedings of the Royal Society B, 2021, 288 (1948): 20210073.

[94] Fei Wu, De-Jun Kong, Peng-Fei Shan, et al. Ongoing green peafowl protection in China[J]. *Zoological Research*, 2019, 40(6): 580-582.

[95] Han L, Liu Y, Han B. The status and distribution of green peafowl *Pavo muticus* in Yunnan Province, China[J]. *International Journal of Galliformes Conservation*, 2007, (1):29-31.

[96] Howard R, Moore A. A complete checklist of the birds of the world[M]. London: MacMillan, 1984:109.

[97] Hurlbert S H. A coefficient of interspecific assciation[J]. *Ecology*, 1969, 50(1):1-9.

[98] IUCN. IUCN Red List of Threatened Species. [EB/OL]. http://www.iucnredlist.org. (2017-06-16).

[99] Jarwadi B H, Ani M, Hadi S A, et al. Behavior Ecology of the Javan Green Peafowl (*Pavo muticus* Linnaeus 1758) in Baluran and Alas Purwo National Park, East Java[J]. *HayatiI Journal of Biosciences*, 2011, 18(4):164-176.

[100] Ke Y Y, Chang H, Zhang G P. Analysis of genetic diversity for wild and captive green peafowl populations by random amplified polymorphic DNA technique[J]. *Journal of Forestry Research*, 2004, 15(3):203-206.

[101] Pudyatmoko S. Habitat and Spatio-Temporal. Interaction between green peafowl with cattle and megaherbivores in Baluran National Park[J]. *Journal of Forest Science*, 2019, 13 (5):28-37.

1 Historical Distribution and Current Situation of Wild Green Peafowl in China

1.1 Distribution of Green Peafowl during Different Historical Periods of China

1.1.1 Discussion on the Distribution of Peafowl in Henan Province

The discovery of peafowl remains (*Pavo* sp.)··· dating from the early Yangshao period at the ninth culture layer of the Xiawanggang dig site in Xichuan County, Henan Province demonstrates that the Yangshao cultural period was an era in which the largest proportion of animals were those that were well acclimated to the heat, and thus the Xiawanggang dig site is representative of the warming period of the time (Jia Lanpo and Zhang Zhenbiao, 1977). The author did not specify in the article whether the peafowl bones found in the excavation were of native or domesticated peafowl, nor whether they were the remains of green or blue peafowl. Wen Huanran and He Yeheng—renowned historical geographers in China—believe that the discovery of the aforementioned peafowl remains indicates that the wild green peafowl was once distributed at the border between the natural forest and open shrubs at the southeast end of the Qinling Mountains five or six thousand years ago (Wen Huanran and He Yeheng, 1980, 1981, 2006; He Yeheng, 1994). Since Henan Province has been regarded as the northern-most distribution range of wild green peafowl in China according to literature thus far, we cautiously and earnestly combed through and analyzed the article, *Animals in the Xiawanggang Site in Xixian County of Henan*

Province in its entirety, and found that besides peafowl remains, other animal remains found at the site reported in the article include those belonging to turtles (*Trionyx* sp.), rhesus macaques (*Macaca mulatta*), dogs (*Canis familiaris*), black bears (*Elenarctos thibetanus*), tigers (*Panthena tigris*), sumatran rhinoceroses (*Didermocerus sumatrensis*), Asian elephants (*Elephas maximus*), wild boars (*Sus scrofa*), domestic pigs (*Sus scrofa domesticus*), Siberian musk deer (*Moschus moschiferus*), muntjacs (*Muntiacus* sp.), and sika deer (*Cervus nippon*), among others. At the end of the article the author emphasizes that the domestic animals of the Yangshao cultural period discovered at the Xiawanggang site, such as pigs and dogs, were very different from their ancestors —wild boars and wolves—indicating that there must have been a long process of evolution before the Yangshao Cultural Period. Based on these materials, we cannot rule out the possibility that the remains of the peafowl (tarsal metatarsals) found in this archaeological discovery are of the domesticated type. We do not have sufficient evidence, to assume this was an area in which wild green peafowl were distributed, making the subject worthy of discussion and further study.

1.1.2 Distribution of Peafowl in Hubei and Hunan Province

Mentions of green peafowl can be found dating back to even some ancient Chinese historical texts. *The Songs of Chu · Dazhao* of the Warring States period refers to a "garden full of peafowl" of which Wang Yi of the Han Dynasty elaborates that, "It refers to a large number of peafowl gathered in the garden." Wen Huanran and He Yeheng (1981, 2006) therefore believe that the peafowl mentioned here are domesticated peafowl, and came to the conclusion that this "reflects that wild peafowl may have been distributed in the Chu State over 2000 years ago, or in the areas surrounding Hubei and Hunan today." After careful analysis, we believe that the presence of domesticated peafowl does not necessarily imply the existence of wild peafowl in the area. This conclusion still warrants discussion; more evidence on the distribution of peafowl is also needed to determine the extent of wild peafowl populations in Hubei, Hunan and other places. He Yeheng (1994) later provides more evidence on the subject. Peafowl are also referenced in the *Changyang County Chronicles · Products* published in the 19th year of Emperor Qianlong (1754) and the 5th year of Emperor Tongzhi (1886) of the Qing Dynasty. In the *Hubei Chronicles · Products* in 1921, it's said that

"peafowl originally came from Changyang, which may be proven by county records (referring to the 1754 and 1886 county chronicles, in which there were also wild peafowl)." In the *Picture Collection · Fangyu Compilation · Zhifang · Yongzhou Prefecture Products Research*, feathered birds include "Mandarin ducks, partridges, golden pheasants, white pheasants, peafowl, parrots," etc. In the early Qing Dynasty, Yongzhou Prefecture had jurisdiction over the counties and cities of Yongzhou, Lingling, Shuangpai, Daoxian, Qiyang, Dong'an, Ningyuan, Jiangshui, Jianghua, and Xintian. Here, peafowl were wild birds like the golden and white pheasants; they were not domesticated. In the *Chenzhou Prefecture Chronicles · Products* made in the 30th year of Emperor Qianlong (1765), the primary wild birds found in the prefecture (now Yuanling, Chenxi, Luxi, Xupu and other counties) included "peafowl, cranes, and partridges." As for the peafowl, the author noted "They are colorful, bright and lovely, with screen-like tails." All of these records indicate that wild peafowl could be found in Hubei and Hunan before the mid–18th century, so the existence of wild peafowl in this area during ancient times may be postulated (He Yeheng, 1994).

1.1.3 Discussion on the Distribution of Peafowl in Zhejiang Province

Before the 18th century, wild peafowl were also present in southwest Zhejiang, a fact that can be proven by the mentioning of peafowl in the *Jinhua Prefecture Yiwu County Chronicles* ·Natural Endowments in the 31st year of the reign of Emperor Kangxi and in the 5th year of the reign of Emperor Yongzheng (He Yeheng, 1994).

1.1.4 Discussion on the Distribution of Peafowl in the Xinjiang Uygur Autonomous Region

In a study of historical documents conducted by Wen Huanran and He Yeheng (1981, 2006), it was discovered that there were also records of peafowl in the Tarim Basin of Xinjiang in northwestern China. In volume 924 of the *Taiping Yulan* it was recorded, "(The Three Kingdoms Period) Emperor Wen of Wei said to the ministers that, the former king of Yutian (now Hotan, Xinjiang) offered ten thousand peafowl tail feathers." *The Kingdom of Qiuci in the Western Regions* from The *Beishi* volume 97 recorded that in Qiuci in the

Northern Wei Dynasty, "A large number of peafowl fly in the valley. People feed and eat them the same as chickens. There are more than a thousand peafowl in the Wang's family" but this has yet to be verified.

Wang Zijin (2013) believes that if the peafowl being referenced to in the Western Regions come from Tianzhu, they prove that Qiuci had a role in the history of transportation between China and foreign countries. In fact, we cannot rule out the possibility that these peafowl were actually native to Quici at that time, since the aforementioned historical documents seem to indicate that they were extremely widespread in the area. Although there are no traces of peafowl in this area today, their existence may be proven by future archaeological work, just like the discovery in Zhechuan. We can also find information that supports this opinion from earlier historical records. According to *Yi Zhou Shu·Wang Hui*, when the feudal princes gathered at Chenghui to meet the emperor, they claimed that the "*Fang* people offered *Kongniao* (peafowl)" (Kong Huang of the Jin Dynasty noted that "*Fang* people refer to the *Rong* people"), this depicts a scene of western tribes offering "*Kongniao*".

Bai Na (2015), based mainly on the *Twenty-Four Histories* and other historical books, classified and analyzed historical documents related to peafowl, and believed that peafowl used to be distributed mainly in today's Yunnan, Guangdong, Guangxi, and Xinjiang, and also in Sichuan, Hunan, Hubei, Hainan and Tibet, and occasionally in Henan on the Central Plains.

1.1.5 Discussion on the Source of Peafowl in China

Wang Yanbo (2018), Wang Yanbo and Guo Fengping (2018) believe that from a global perspective, peafowl in China came largely from neighboring areas. They maintain that these peafowl came from two areas. The first is from Southeast Asia, including the Indochina Peninsula, etc. They arrived in China's Yunnan, Guizhou, Guangdong and Guangxi via land or through waterways leading from the Indochina Peninsula, and were brought to the Central Plains via merchants or by tribute. During the Qin and Han dynasties, the green peafowl that inhabited Southeast Asia had already settled in the south of the Five Ridges in China. However, since their numbers were still small, they were offered to the royal family or nobility as rare and exotic animals. According to historical records, Zhuang Qiao, a General from the same time period as Wei Tuo, entered Tian from the Chu State during the Qin and Han Dynasties. Tian is today's Yunnan-Guizhou region, which bordered the Indochina Peninsula and conducted exchange and communication with people in the peninsula for a very long time throughout history. The peafowl mentioned here were either native species or brought from southern islands. The other area from which peafowl arrived in China is the Indian Peninsula and Sri Lanka. They travelled to Central Asia via Southern Asia, crossed the Congling Mountains through Xinjiang as far as the Hexi Corridor along the ancient Silk Road, and arrived in the Central Plains. The peafowl here seemed to have existed in large numbers in the Western Regions in the Middle Ages. However, we cannot confirm whether the peafowl here were Indian blue peafowl or Javan green peafowl. Based on insights into ideal living conditions for peafowl, it seems that the Western Regions in which Qiuci was located would have been unable to provide either the climate or the geographic requirements for peafowl to thrive. The blue peafowl, who have better adaptability and are more suitable to the conditions on the route, were likely the peafowl mentioned in this area.

After consulting a large volume of zoological literature, we believe that:

① The conclusion of this study that peafowl in China came from neighboring areas, and those in southwestern Yunnan, Guangdong and Guangxi originated from the southeast and Indochina Peninsula is wrong. The existence of many native green peafowl in many areas in Yunnan proves the idea.

② Since there is no description in ancient texts about whether the peafowl are blue or green peafowl, it's reasonable in this article to come to the conclusion that the peafowl in the Western Regions/Xinjiang were likely blue peafowl, which came from the Indian Peninsula and Sri Lanka, based on the habitat requirements of the green peafowl.

1.1.6 Historical Distribution of Wild Green Peafowl in China

Wen Huanran et al. (1981, 2006) and He Yeheng (1994) believed that peafowl present in historical periods in China were mainly distributed over the Yangtze River Basin and to the areas south of it. They can be roughly divided into the following three regions: the

Yangtze River Basin, south of the Five Ridges, and southwestern Yunnan. The distributionis is described as follows:

(1) The Yangtze River Basin includes northeastern Yunnan, the Sichuan Basin, and the middle reaches of the Yangtze River

The Sichuan Basin in the upper reaches of the Yangtze River and northeastern Yunnan were places most frequented by peafowl in ancient times. Zuo Si of the Jin Dynasty points out in his *Ode to the Capital of Shu* that "peafowl and kingfishers flock, and rhinos and elephants race" in the Sichuan Basin. Liuliang of the Tang Dynasty noted, "*Kongniao* refers to the peafowl while *Cuiniao* refers to the kingfisher." This shows that there were many wild peafowl in this area at the end of the 3rd century.

In *The Book of the Later Han Dynasty · Biographies in Southern and Southwestern Regions · Dian*, it's mentioned that the "Yizhou Prefecture is flat and vast. There are parrots and peafowl. It is rich with salt, ponds, fields, fish, gold, silver and livestock." Chang Qu of the Jin Dynasty said in his *Huayang Guozhi · Nanzhongzhi*, "Jinning Prefecture refers to Yizhou. It has jurisdiction over Dianchi Lake (now Jinning Jincheng, Yunnan). It covers a vast and abundant land. There are parrots, peafowl, salt ponds, fields and fish." These materials show that more than 1,000 years ago, there were many wild peafowl in the northeastern part of Yunnan (He Yeheng, 1994). Most in need of explanation in these documents is that modern-day Jinning is a district of Kunming City. It is located to the south of the main urban area of Kunming and belongs to the central Yunnan region instead of northeastern Yunnan.

(2) South of the Five Ridges

According to ancient records, peafowl were mainly distributed in the following six regions geographically: eastern Guangdong (Chaoyang County, Guangdong Province), Central Guangdong (Conghua County, Guangdong Province), Yunkai Mountain and its surrounding areas (Xuwen County, Lianjiang County, Haikang County, Wuchuan County, and Gaozhou of Guangdong Province, as well as Yulin County and Bobai County of Guangxi Zhuang Autonomous Region), Northern Guangxi (Mengshan County, Yishan County, Lingui County, Xing'an County, Lingchuan County, Yongfu County, and Yangshuo County of the Guangxi Zhuang Autonomous Region), Southwest Guangxi (Wuming County, Yizhou County, Yujiang County, Hongshui River, Zuojiang, Youjiang, Yujiang and the Qianjiang River Basins in around the Guangxi Zhuang Autonomous

Region), and the Southeast Guangxi Regions (from Liuwan Mountain to Shiwan Mountain in the southern area of the Guangxi Zhuang Autonomous Region, namely Hepu County, Qin County, Shiluodu) (Wen Huanran et al., 1981, 2006). From the Western Han Dynasty to the Southern Song Dynasty, peafowl were frequently seen to the south of the Five Ridges (He Yeheng, 1994). There are many peafowl in the western mountains of the Guangxi Zhuang Autonomous Region. According to the *Trip in Western Guangxi* of the Qing Dynasty, there were many wild peafowl in Nanning, Laibin, Guiping, and in the Hejiang river basin in the Guangxi Zhuang Autonomous Region (Wen Huanran and He Yeheng, 1980). In *On Salt and Iron*, it was said that "There is peafowl bait at the doors in Southern Guangdong" and *Hao Lu · Geography* also wrote that "There are a large number of peafowl in the prefecture," showing that Hainan Island was also home to peafowl (Zeng Zhaoxuan, 1980).

(3) Southwestern Yunnan

Southuestern Yunnan is an area with a long history of wild peafowl in China. According to *Huayang Guozhi · Nanzhongzhi*, "Yongchang Prefecture, or the ancient Ailao State, has fertile land and rich products," and that "there are peafowl, rhinos, elephants, and other rare birds and animals." Ailao is an ancient state. In the 12th year of Yongping in the Eastern Han Dynasty (AD 69), Ailao (now east of Yingjiang County, Yunnan) and Bonan (now Yongping County) were under the jurisdiction of Yongchang Prefecture. At that time, Yongchang Prefecture had a wide jurisdiction, covering more than 3,000 Li from east to west and 4,800 Li (1Li=500m) from north to south, roughly the same as the present day Dali Bai Autonomous Prefecture, Baoshan City, Lincang City, Dehong Dai, the Jingpo Autonomous Prefecture, and the Xishuangbanna Dai Autonomous Prefecture, etc. Since the Han Dynasty, southwestern Yunnan has been famous for its peafowl. For example, in the Tang Dynasty, peafowl could be commonly found in the Dehong Dai and Jingpo Autonomous Prefecture of Yunnan Province, and it's reported that many people had peafowl resting in the trees in front of and behind their houses. This indicates that peafowl must have had a large presence at the time (Wen Huanran and He Yeheng, 1980, 2006). Wild peafowl were even present from the Han Dynasty and the Yuan Dynasty (Wen Huanran and He Yeheng, 1980, 1981, 2006). According to the *Yunnan Chronicles* in the 6th year of Emperor Longqing (1572) and the *Dian Chronicles* in the 5th year of Emperor Tianqi (1652) of the Ming Dynasty, peafowl were distributed in Wenshan,

Kaiyuan, Kunming, Fumin, Chuxiong and other places. Wild green peafowl were widely distributed in Jianshui County, Dali City, Baoshan City, Lianghe County and Yuanjiang County, Zhenyuan County, Jingdong County, Fengqing County as well as other places (He Yeheng, 1994). In the Qing Dynasty, there were even records detailing the distribution of wild green peafowl in Kunming City, Chuxiong City, Yuxi City (Yuanjiang Hani, Yi and Dai Autonomous County), Pu'er City (Zhenyuan County, the Jingdong Yi Autonomous County, Ning'er Hani, the Yi Autonomous County), the Dali Bai Autonomous Prefecture (Dali City, Yong Ping County), Lincang City (Fengqing County), Baoshan City (Longyang District, Tengchong City), Honghe Hani, the Yi Autonomous Prefecture (Jianshui County), the Xishuangbanna Dai Autonomous Prefecture (Jinghong City), Dehong Dai, and the Jingpo Autonomous Prefecture (Mangshi, Lianghe County, Longchuan County).

According to the *Distribution and Migration Map of Peafowl in China* (Fig.1–1) drawn by Wen Huanran and He Yeheng, these distribution locations include counties, cities and districts in Guangdong Province, the Guangxi Zhuang Autonomous Region, and Yunnan Province. The specific locations are as follows:

Guangdong Province: Guangzhou City, Chaoyang County (now Chaoyang District, Shantou City), Luoding County (now Luoding City), northeast of Gaozhou County (now Gaozhou City), southwest of Wuchuan County (now Wuchuan City), north Lianjiang County (now Lian Jiang City), Haikang County (now Leizhou City), and Xuwen County.

The Guangxi Zhuang Autonomous Region: Nanning City, Wuming County, Shanglin County, south of Mengshan County, northwest of Yulin City (now Yuzhou District of Yulin City), Rong County, Bobai County, Qinzhou City, Lingshan County, Hepu County, Fangcheng Autonomous counties (now Fangchenggang City), Chongzuo City (now Jiangzhou District, Chongzuo City), Yishan County (Yizhou City).

Yunnan Province: Jinning County, Yuanjiang Hani, the Yi and Dai Autonomous County, Baoshan City (now Longyang District, Baoshan City), Pu'er (now Ning'er Hani and the Yi Autonomous County), the Jingdong Yi Autonomous County, Zhenyuan Yi, the Hani and Lahu Autonomous County, Fengqing County, Lianghe County, Yingjiang County, Longchuan County, Yongping County, and Jinghong County (now Jinghong City).

▲ Fig.1-1 Distribution and Migration Map of Peafowl in China (quoted from Wen Huanran et al., 2006, revised)
The map boundary in this figure doesn't constitute the basis for defining different areas

In addition to the areas in southwestern Yunnan mentioned in the distribution of wild green peafowl in China during the above historical periods, peafowl were also distributed in other regions of Yunnan Province. For example, Bai Na (2015) believes that peafowl were also distributed throughout central Yunnan, western Yunnan, eastern Yunnan, and southern Yunnan.

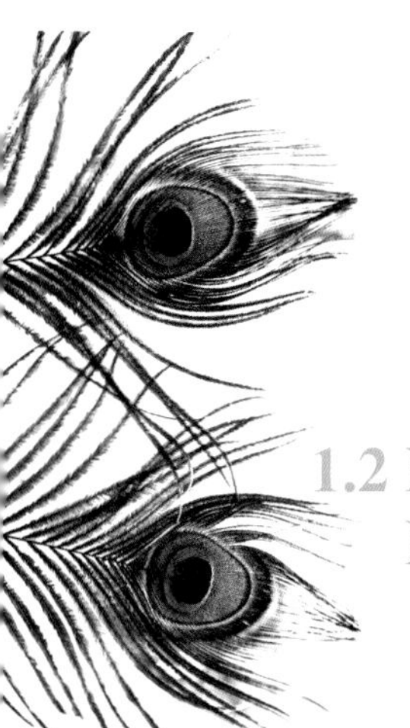

1.2 Historical Regression of Distribution Areas and Regional Extinction of Wild Green Peafowl in China

1.2.1 Historical Regression of Distribution Areas of China's Wild Green Peafowl

Lan Yong (2002) believes that the distribution of wild peafowl in China moved 7.7° southward from 33.1° N latitude in southern Henan to 25.4° N during different historical periods. Wen Huanran and He Yeheng (1981, 1995, 2006) believe that the geographical distribution areas of Chinese peafowl throughout history have gradually moved from north to south and from northeast to southwest. At present, southwestern Yunnan has become the only distribution area of wild peafowl in China. The Yangtze River Basin was the first to be developed, then Pearl River Basin, and finally southwestern Yunnan. This has moved the northern boundary for green peafowl southward, similar to the northern-most distribution boundaries of elephants and rhinos. Wang Yanbo and Guo Fengping (2018) believe that through analyzing historical documents from the perspective of environmental history, peafowl in China have been migrating from north to south and finally southwest.

1.2.2 Historical Regression of Distribution Areas of China's Wild Green Peafowl in the South of the Five Ridges

Domestic scholars have a considerable understanding of the regional changes in the main distribution areas of peafowl in China. As Wen Huanran and He Yeheng mentioned in their research papers and monographs, the extinction of peafowl in the Yunkai Mountain and its surrounding areas in the northeast seemed to take place after the 1830s. In northern Guangxi, there were many peafowl in the mountains in Yizhou (now Yishan County) during the Song Dynasty, but have rarely been seen in any records ever since. After the 18th century, wild peafowl numbers in southeast Guangxi gradually decreased. In summary, wild peafowl were widely distributed to the south of the Five Ridges during different historical periods. They became extinct first in the north and east, and then in the south and west. In other words, they appeared in the north and central part of Guangxi and Guangdong first, then in the south, initially in Guangdong and then in Guangxi, and then in the plains before moving to the mountains. The wild peafowl in Lingshan County between the Liuwan and Shiwan mountains in southeastern Guangxi did not become extinct until the 20th century. Shifting trends in the extinct of wild peafowl in the south of the Five Ridges also roughly reflects the general trend of regional development (Wen Huanran and He Yeheng, 1981, 1995, 2006).

1.2.3 Distribution Records of Peafowl in the Pearl River Delta

Peafowl have also been widely distributed across southern China. For example, it's written in the *Nanyue Chronicles* that "There are many peafowl in Dushan, Yining County" and in the *Anecdotes in the South of the Five Ridges* of the Tang Dynasty it was also recorded that there were peafowl "burdened with green tails." From this we can conclude that there were peafowl in Guangzhou at that time. In *Beihulu*, "There are peafowl in the ten prefectures, including the Nan, Xin, Qin, Chun prefectures." We can see that there were peafowl inhabiting Nanhai and Xinhui in the Tang Dynasty just as in the mountainous areas of western Guangdong. As Yining is now Kaiping, we can see that the mountainous areas around the Pearl River Delta have been peafowl distribution areas since the Jin Dynasty. There were still many peafowl in these areas in the Tang Dynasty. For example, in the *Xinhui Chronicles* (1908), "In the Zhenyuan period (785–850) of the Tang Dynasty, Dugu was the governor of Guangzhou. He said Xinhui Guishan had emeralds, peafowl, and apes while under his jurisdiction. There were rare birds in the surrounding areas of Guifeng Peak. Other mountains and forests were not yet fully opened." After the Song Dynasty, however, there were few records in the delta area. The massive hunting of peafowl in the Southern Song Dynasty was recorded in *Ling Chu Dai Da*. In volume 9 of this book it's said, "When Zhongzhou people got one peafowl, they would exchange it for coins. The southerners would preserve and eat it. The peafowl were cheap for the local people." After the Song Dynasty, peafowl were rarely seen in the records in the Pearl River Delta region. Therefore, the peafowl in the delta area became extinct earlier than those in Guangxi and western Guangdong. There were still records in Guangxi and western Guangdong in the works in the Qing Dynasty. For example, in the *Nanyue Notes*, "Peafowl originated in Guangxi, and there may be peafowl in the Luoding Mountain." We can also see that the breeding area in Guangdong had been greatly reduced.

1.2.4 Historical Regression of Wild Green Peafowl in Southwest Yunnan

Wild peafowl inhabited southwestern Yunnan from the Han Dynasty to the Yuan Dynasty. The distribution of wild peafowl in this area during the Ming and Qing Dynasties was generally limited to the south and west of Yuanjiang, Zhenyuan, Jingdong, Fengqing, and Baoshan. Peafowl were produced in counties such as Zhenyuan, Jingdong, etc. in the Qing Dynasty, but they are now extinct. We can draw the conclusion that the distribution range of wild peafowl in southwestern Yunnan has been further reduced (Wen Huanran and He Yeheng, 1981, 1995, 2006).

1.3 Reasons for Distribution Changes of Wild Green Peafowl in China

Wen Huanran, He Yeheng (1980, 1981, 2006), and He Yeheng (1994) believe that the distribution range of Chinese peafowl rapidly reduced throughout history, and the number of peafowl decreased quickly. There may be various reasons for this. The main reason is human hunting, which was conducted for three reasons. The first was for food. As early as the Tang and Song dynasties, peafowl were hunted in large numbers in the south of the Five Ridges for their meat. It has even been recorded in historical documents that "people in the valley cook and eat [green peafowl]." Secondly, it seems that peafowl were also hunted for decoration. Peafowl feathers and tail feathers are dazzlingly beautiful, and the Emperor Cao Pi of the Wei State of the Three Kingdoms period used them as a cart cover, while Nanyue in the Western Han Dynasty used them for decorating doorways. People in ancient times used them to decorate boat awnings and fan brushes. Thirdly, peafowl were hunted for medicine. Shenhui of the Tang and Song Dynasties quoted *Hua Zi Zhu Jia Ben Cao* in volume 19 of his *Chong Xiu Zheng He Ben Cao* that peafowl could be used for medicinal purposes. The unrestricted hunting of peafowl throughout history and the continuous reclamation of mountains, forests, and grasslands, have destroyed the habitats of peafowl. As a result, the number of peafowl rapidly declined, and the range of their distribution narrowed. The changes in the number of peafowl over the course of history are a reflection of the changes in both people and the biosphere. Human activities, especially the mass hunting of peafowl, are the main reason for these changes.

Wang Yanbo and Guo Fengping (2018) believe that the internal asymmetrical interaction of the entire ecosystem is also a reason for the drastic reduction in the population of peafowl and their retreat to Yunnan. In fact, humans, animals, and nature exist as an organic whole. However, human beings, since their birth, have continuously expanded their living space with the technological tools derived from various experiences in nature. This means that the ecosystem is constantly being affected by human activities, resulting in uncertain impacts. This has minimized and even infringed on animals living in nature that are almost equal to humans. Climate, vegetation, mountains, and rivers have all been eroded in the process of human population expansion. These, however, are precisely the conditions for the survival of animals such as peafowl that act as environmental indicators.

Dong Feng et al. (2021) showed that human interference was the biggest influencing factor since neolithic age.

1.4 Current Distribution and Population of Wild Green Peafowl in China

1.4.1 Investigation Reports on the Distribution of Wild Green Peafowl in China in the 20th Century

(1) Survey on the distribution and population of wild green peafowls in China

In the spring and summer of 1960, Kuang Bangyu primarily observed the ecology and hunting methods for the peafowl in the Xishuangbanna Dai Autonomous Prefecture, Mojiang County in the Simao Prefecture (now Pu'er City), and Xinping County (namely Xinping Yi and Dai Autonomous County) in the Yuxi Prefecture (now Yuxi City) in southern Yunnan. According to the survey results, he came to the conclusion that domestic peafowl are only found in southern Yunnan, especially in Simao (now Pu'er City), Yuxi (Yuxi City), Xishuangbanna (now the Xishuangbanna Dai Autonomous Prefecture), Lincang (now Lincang City), Dehong (now the Dai Jingpo Autonomous Prefecture) among other places (Kuang Bangyu, 1963).

Wen Huanran and He Yeheng (1980) believe that according to surveys collected from animal workers, there are wild peafowl in Mengzi County in Honghe Hani and the Yi Autonomous Prefecture, the Hekou Yao Autonomous County, the Simao region, Menghai

County in the Xishuangbanna Dai Autonomous Prefecture, Lincang County in the Lincang region, Yingjiang County in the Dehong Prefecture and Hushui County in the Nujiang Lisu Autonomous Prefecture.

Wang Zijiang (1990) reported that traces of peafowl were found on the Chuxiong Zixishan Nature Reserve in 1987. On January 20, 1989, Wang Zhaoming from the Management Office of the Baizhushan Nature Reserve, Shuangbai County, saw 4 peafowl (1 male, 3 female). During his visit to the Chuxiong Yi Autonomous Prefecture in Yunnan from February 14 to March 6, he heard the calls of peafowl as early as 10 a.m. on February 26, which were confirmed to have indeed been peafowl (*Pavo muticus imperator*). According to extensive investigations, there are peafowl in Lufeng County, Shuangbai County, Nanhua County, and Yao'an County in Chuxiong Prefecture, as well as in Chuxiong City.

He Yeheng (1994) believes that southwestern Yunnan has become the only distribution area for wild peafowl in China. According to the investigations and specimens captured by modern animal workers, peafowl in China at present are only distributed in the Honghe Hani and the Yi Autonomous Prefecture (Mengzi County, north of Hekou Yao Autonomous County), the Simao region (now Pu'er City, editor's note), the Xishuangbanna Dai Autonomous Prefecture, Lincang City, the Dehong Dai and Jingpo Autonomous Prefecture, and the Nujiang Lili Autonomous Prefecture (east of Hushui County) in southwestern Yunnan. Although Hushui and other places were not recorded in the Qing Dynasty literature, Yuanjiang, Zhen Yuan, Jingdong, Dali, and Baoshan, among others, all documented the presence of wild peafowl in the Qing Dynasty which are now extinct. The distribution range of wild peafowl in southwestern Yunnan has been progressively reduced. Looking at the literature above, the geographical distribution of wild peafowl in China throughout history has gradually narrowed from north to south, and from northeast to southwest.

Wen Xianji et al. conducted letter surveys and selective field investigations from 1991 to 1993. They believe that the regions with large population of peafowl are as follows: Ruili County (Ruili City now) (40–50), Longchuan County (70–90), Changning County (30–40), Yongde County (30–50), Xinping County (40–60), Pu'er County (now Ning'er County, 30–40), Mojiang County (30–50), Jingdong County (30–40), Chuxiong City (50–80), Shuangbai County (150–250), Nanhua County (50–100), Luxi County (now Mang City, numbers

unknown), Longling County (10–20), Yun County (15–30), Lincang County (now Linxiang District, 15–20), Fengqing County (5–10), Shuangjiang County (numbers unknown), Cangyuan County (20–30), Zhenkang County (numbers unknown), Gengma County (numbers unknown), Jinggu County (20–30), Yao'an County (numbers unknown), Yongren County (numbers unknown), Lufeng County (numbers unknown), Weixi County (numbers to be confirmed), Deqin County (numbers to be confirmed), Jinghong County (now Jinghong City, numbers unknown), Menghai County (numbers unknown), Mengla County (numbers unknown), Weishan County (numbers unknown), Yingjiang County (may be extinct), Lushui County (may be extinct), and Tengchong County (now Tengchong City, extinct). Based on this, they believe that green peafowl are currently only seen in the western, central, and southern parts of Yunnan in China. Areas that once possessed distribution records but in which peafowl are now extinct or endangered include Yingjiang County, Lushui County, Tengchong County, Mengzi County, and Hekou County. Zhonghe and Zhizuo in Yongren County are discovered distribution areas at that time. According to local residents, green peafowl have also been found in Yezhi in Weixi County, as well as in Tuoding and Benzilan in Deqin County. Due to habitat destruction, the existing population of green peafowl have scattered into small family groups. The estimated number of existing populations reported by all the counties is cumulatively 635 to 950. Since areas with unknown distribution numbers are not included, it is estimated that the number of existing wild peafowl in China is about 800 to 1,100 (Wen Xianji et al., 1995). This survey covers most areas of Yunnan Province. It can be considered as the first comprehensive and systematic survey of wild green peafowl in China. Although it provides an estimated population, it is of vital importance to estimate the current status of wild green peafowl in China. Furthermore, the article also drew a distribution map of green peafowl in China in the 1980s (Fig.1–2).

In this research report, Wen Xianji et al. also claim that green peafowl had once been distributed throughout Hunan, Hubei, Sichuan, Guangdong, Guangxi, and Yunnan provinces at various points in history. By the beginning of the 20th century, they were extinct in the other provinces and in northeastern Yunnan, and their distribution area had shrunk to the western, central, and southern parts of Yunnan Province (Wen Xianji et al. 1995). This is more accurate than previous ideas in which wild green peafowl in Yunnan Province were only distributed in northeastern Yunnan or southwestern Yunnan. This is important for being able to put a solid grasp on the distribution area of green peafowl in Yunnan and China.

Yang Lan et al. (1995) recorded the geographical distribution of green peafowl in Yunnan Province county by county: Xinping, Pu'er (now Ning'er), Jingdong, Jinggu, Simao, Mojiang, Menglian, Lincang (now Linxiang), Yongde, Fengqing, Cangyuan, Zhenkang, Gengma, Mile, Shiping, Mengzi, Jinping, Hekou, Luchun, Menghai, Chuxiong, Lufeng, Shuangbai, Nanhua, Yao'an, Luxi (now Mang City), Longchuan, Yingjiang, and Lushui, among others. These counties are essentially the same as those reported by Wenxianji et al. (1995).

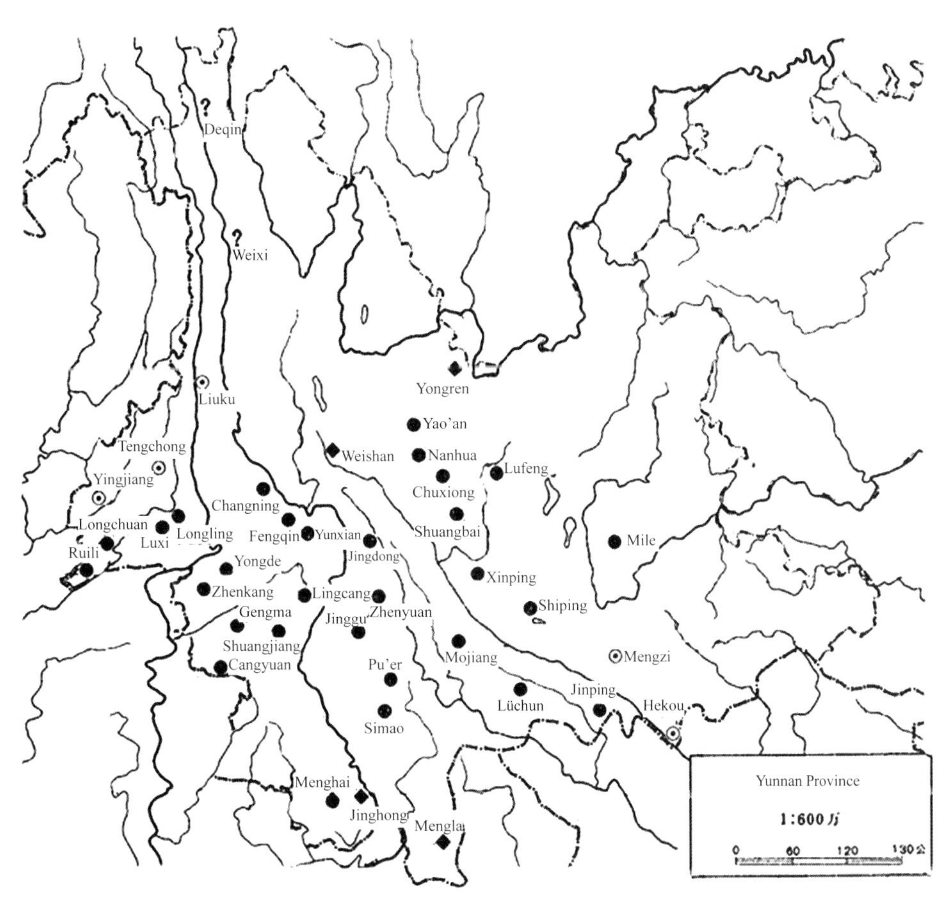

▲ Fig.1–2 Distribution of Green Peafowl in China in the 1980s (Wenxianji et al., 1995)
The map boundary in this figure doesn't constitute the basis for defining different areas

(2) Distribution and quantity of wild green peafowl in some areas

① Chuxiong Yi Autonomous Prefecture.From February 1990 to February 1995, Xu

Hui explored the counties, townships, villages, nature reserves and forest areas where green peafowl might be distributed in the Chuxiong Yi Autonomous Prefecture, Yunnan Province, and also interviewed local hunters, farmers, township officials, employees in natural conservation and expansion arcas, and seniors in the prefecture who have conducted in-depth research on wildlife. Through personal observation, investigation, and interview, he arrived at a deeper understanding of the distribution, quantity, and habitual behaviors of peafowl in the Chuxiong Prefecture. According to inspections and surveys, and after consulting with the relevant information, he concluded: Green peafowl in the Chuxiong Yi Autonomous Prefecture are distributed only in 29 villages/towns in five counties/cities including Chuxiong City (which encompasses Qianjin Township, Zhongyishe Township, Dadiji Township, Lvhe Township, Donghua Township, Xi Shelu Township, Duoju Township, Xincun Township, Yunlong Township, Muchong Township), Shuangbai County (Ejia Town, Taihejiang Township, Aini Township, Tuodian Town, Yulong Township), Yao'an County (Mixing Township, Dahekou Township, Bala Township, Zuomen Township, Guantun Township), Nanhua County (Majie Township, Xuying Township, Shaqiao Township, Tianshentang Township, Yulu Township, Longchuan Township), Lufeng County (Jiuzhuang Township, Nanhe Township and Luochuan Township). The number of wild green peafowl in the prefecture is 280 (including records of 19 that were poisoned, 31 hunted, and 23 captured), among which, there are 104 in Chuxiong City, 79 in Shuangbai County, 35 in Lufeng County, 30 in Nanhua County, and 32 in Yao'an County. In these five counties (cities), green peafowl are distributed in the following nature reserves: 2 in Chuxiong City (Ailaoshan National Nature Reserve Chuxiong Area, Chuxiong Zixishan Nature Reserve), 2 in Shuangbai County (Ailaoshan National Nature Reserve Shuangbai Area, Shuangbai County, Baizhushan Nature Reserve), one in Nanhua County (Nanhua County Dazhongshan Nature Reserve), and 1 in Lufeng County (Lufeng County, Zhangmuqing Nature Reserve) (Xu Hui, 1995).

② Lincang City.From July to November 1992, Guo Baoyong et al. (1999) conducted surveys on wildlife in the Nangunhe National Nature Reserve in Yunnan using the interview survey method and the line transect method. Six green peafowl were encountered. It's estimated that there are 48 green peafowl in the nature reserve.

③ Xishuangbanna Dai Autonomous Prefecture.In 1994, Luo Aidong and Dong

Yonghua conducted a survey on the number and distribution of wild green peafowl in the Xishuangbanna Dai Autonomous Prefecture through a combination of visits and via "listening stations" to collect sound-based statistics. The results showed that there are 19–25 green peafowl in Xishuangbanna. (Including 3 peafowl, one male and two females, the rosin-cutters seen in the pine wool forest in the Zhengnuo Township, Jinghong City), which account only for 2.3%–2.4% of the total number of Chinese green peafowl (800–1100). If calculated according to an area of nearly 1.969×10^4 km^2 in Xishuangbanna, the population density of green peafowl is only $(0.95–1.25) \times 10^{-3}$ per km^2, which is far lower than the provincial average of $(2.1–2.9) \times 10^3$ per km^2. Green peafowl are already endangered and require urgent protection (Luo Aidong and Dong Yonghua, 1999). According to the survey data, Luo Aidong and others also analyzed the reduction rate of the green peafowl distribution area in the Xishuangbanna area. From 1980 to 1985, the distribution area of wild green peafowl was 118 km^2. From 1990 to 1995, the distribution area was 23 km^2. From 1985 to 1990, the distribution area was only 18.7 km^2. Compared with the 1980s, the reduction rate of the distribution area is 84.15%. They are only distributed in three narrow regions, including the Dadugang Township and Jingne Township in Menghai County, and the Mangao Nature Reserve in the Xishuangbanna National Nature Reserve in Menghai Town, Menghai County. The existing population distribution areas are isolated from each other, and there is no communication between the populations, seriously hindering efforts to increase the population and promote the development of the species (Luo Aidong and Dong Yonghua, 1999).

④Southeast Yunnan.According to the survey results in 1995 and 1996, Yang Xiaojun et al. (1997) believe that green peafowl are distributed only in Jianshui, Shiping and Mile counties in southeastern Yunnan, which is consistent with the locations described by Zheng Baolai et al. (1987) and Zhang Fan et al. (1987). It is estimated that there are about 50 green peafowl in southeastern Yunnan.

Jianshui County: There are about 20 in Guanting and Qinglong townships, less than five in Potou Township, and 10~15 in Panjiang Township. Limin Township also reports the presence of peafowl but the exact number is unclear.

Shiping County: Similarly, Longpeng and Longwu townships also possess green peafowl, but the exact population is unknown. There are less than five in Baoxiu Township.

Mile County: Green peafowl are occasionally seen by the Inspection Department.

Green peafowl in six counties and cities including Mengzi, Jinping, Luchun, Hekou, Kaiyuan and Wenshan are extinct. Moreover, green peafowl were seen in Gejiu City in the early 1980s, but are now extinct.

⑤ Northwest Yunnan.Two green peafowl were found along the Zhibalo River in Deqin County, northwestern Yunnan Province in 1986. For this reason, the villages in question have formulated a rule for villagers to protect these two green peafowl. A green peafowl was later captured on March 15, 1986 in Duosong Village, Xiaruo Township (2200m above sea level), and the hunter was fined and detained by the Nature Reserve Police Station of Deqin County Public Security Bureau. The green peafowl skin was confiscated by the Baima Snow Mountain Nature Reserve Management Bureau. Since then, no green peafowl has been found. Based on this and according to the survey results in 1995 and 1996, Yang Xiaojun et al. (1997) believe that the sightings of green peafowl in the area may only be an occasional phenomenon.

1.4.2 Investigation Reports on the Distribution of Wild Green Peafowl in China in the 21st Century

(1) Distribution and population of wild green peafowl in China

According to Wen Yunyan et al. (2016), a survey conducted in 2013–2014 by the Kunming Animal Research Institute of the Chinese Academy of Sciences show that in the past five years, only 14 sites in 11 out of the 34 counties and cities with historical peafowl distribution have field records of the green peafowl, where the population of the green peafowl has been estimated to be less than 500, meaning that the green peafowl is now one of the most endangered wildlife species in China.

As the editor-in-chief of *Classification and Distribution Directory of Chinese Birds (3rd Edition)*, Academician Zheng Guangmei expresses his believe in the text that in China, green peafowl are distributed in Southeast Tibet and Yunnan (Zheng Guangmei, 2017).

Yang Xiaojun et al. (2017) report that they conducted research on green peafowl in China from April 2014 to June 2017, using questionnaires, interviews, and path methods, and supplemented the research by using plotting, bird calls, and infrared cameras, among other methods, while also analyzing the relevant literature. The results show that 52 counties in China once recorded the presence of green peafowl, but only 23 counties still possess any green peafowl population. Compared with results based on the research of Wen Xianji and his colleagues in 1991–1993, the population has been on a sharp downward curve, from 800–1100 to less than 500.

During 2015–2017, the line-transect and point-transect methods were used combined with interview investigations to conduct research into the population and current distribution of green peafowl by researchers led by Rong Hua. Relevant areas were selected to carry out correspondence and field investigations, including Yuxi, Chuxiong, Pu'er, Xishuangbanna, Lincang, Baoshan, Dehong, Nujiang, Diqing, Lijiang, Dali and other cities/prefectures in Yunnan Province. The results show that the current population of wild green peafowls in China is significantly reduced from 800–1100 two decades ago to about 235–280 (Wen Xianji et al., 1995), while the distribution area sharply reduced from 32 counties in Yunnan Province in 1995 to 13 counties, and in these 13 counties the actual population area has been limited to some coastal areas of the Longling and Yongde section of the Nujiang River Basin, the Jinggu section of the Lancang River basin, and the Shiyang and Lüzhi rivers of the Honghe River basin. The estimated number of green peafowl for each county is given: 60 in the Xinping Yi Dai Autonomous County (Yuxi City); 40–50 in Shuangbai County (the Chuxiong Yi Autonomous Prefecture); 20–30 in Longling County (Baoshan City); Linxiang District (Lincang City) (unknown, pending further investigation); Yun County (unknown number, pending further investigation); Fengqing County (unknown number, pending further investigation); 30–40 for Yongde County; 25–30 for Zhenkang County; Gengma Dai Wa Autonomous County (unknown number); Cangyuan Wa Autonomous County (unknown number); Shuangjiang Lahu Wa Blang Dai Autonomous County (unknown number); 30 in the Jinggu Dai Yi Autonomous County (Pu'er City); 30–40 in Ruili City (Dehong Dai Jingpo Autonomous Prefecture). Furthermore, the results show that in Yunnan, China, green peafowl mainly inhabit tropical and subtropical areas, especially in forests with evergreen and broad-leaved trees along the banks of valleys and in relatively open and low-density pine forests. Due to habitat destruction, the existing population of the green peafowl has been scattered across their distribution areas in small family clusters. (Rong Hua et al., 2018).

With a thorough inquiry of the pertinent literature as a backup, Kong et al. (2018) conducted a field investigation in 11 cities in Yunnan Province and 2 prefectures in the Tibet Autonomous Region where wild peafowl may have had a presence. Simultaneously, 190 line transects (784 km) were put into effect in 24 counties in Yunnan and 19 line transects (81 km) in the Tibet Autonomous Region from 2014 to 2017. The results show that in the past 30 years, the distribution of the green peafowl has been recorded in 52 counties of 11 cities in Yunnan Province and 2 counties in 1 city in the Tibet Autonomous Region. However, over the past 20 years, from 1991 to 2000, the distribution range has been reduced from 127 towns in 34 counties in 11 cities to 33 towns in 22 counties in 8 cities (Fig.1–3), which means 35 percent of the counties and 74 percent of the towns have lost their green peafowl in the last 20 years. In this study, no green peafowl were found in Motuo and Chayu counties of the Tibet Autonomous Region. The Yunnan Chuxiong Yi Autonomous Prefecture, Yuxi City, Pu'er City, Lincang City, Baoshan City, the Dehong Dai Jingpo Autonomous Prefecture, the Honghe Hani Yi Autonomous Prefecture, and the Xishuangbanna Dai Autonomous Prefecture are located in central, western, and southern Yunnan, as well as eight other prefectures or cities. In the Dali Bai Autonomous Prefecture, the Diqing Tibetan Autonomous Prefecture, and the Nujiang Lisu Autonomous Prefecture, there is no iron-clad evidence to support the existence of the green peafowl. In the Yuanjiang and Eshan areas in central Yunnan, green peafowl that were not previously recorded have been found. Based on this study, *Kong* et al. found a sharp population decline compared to 20 years ago. For example, according to results of the 1991–2000 investigation, 585–674 green peafowls were recorded in 34 counties in Yunnan Province, while in this study, using the same interview methods, there were only 183–240 green peafowls recorded in 48 counties. Of the 34 counties with green peafowl distribution records in the 1991–2000 investigation, 30 counties show a downward trend in the population of the green peafowl, while the population of the green peafowl in the two counties adjacent to Shuangbai County and Xinping County in central Yunnan Province have increased significantly, accounting for more than 60% (63.93%–69.17%) of the total population of the peafowl (according to the interview results) in Yunnan Province.

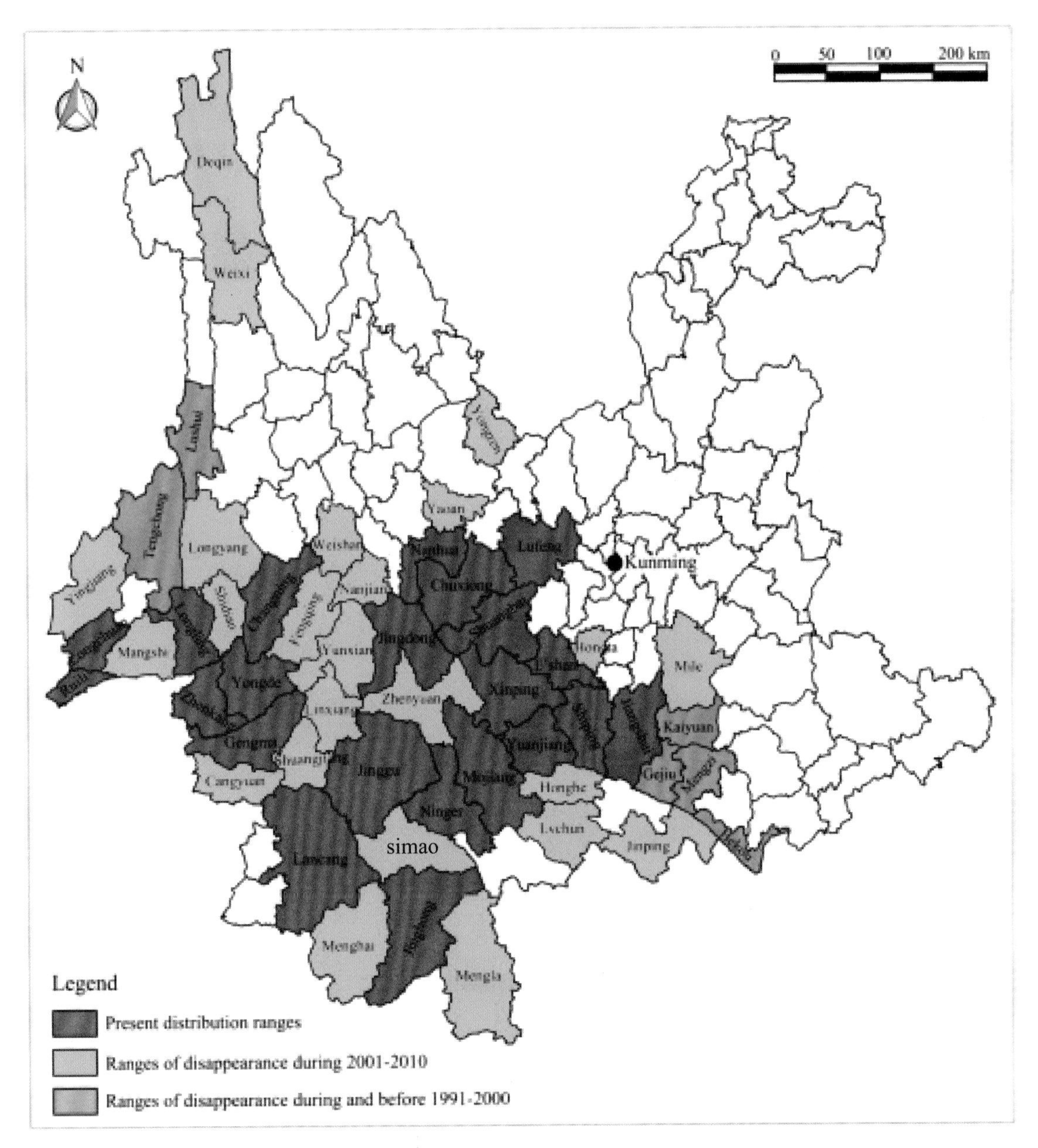

▲ Fig.1–3 Present Distribution of Green Peafowl in China (From Kong et al, 2018 revised)
The map boundary in this figure doesn't constitute the basis for defining different areas

Wu et al. (2019) provide the distribution map of the wild green peafowl and population size in Yunnan, China (Fig.1–4). It also mentions that the main challenge for protecting the green peafowl is that more than 65% of green peafowl in China inhabit areas outside of conservation zones.

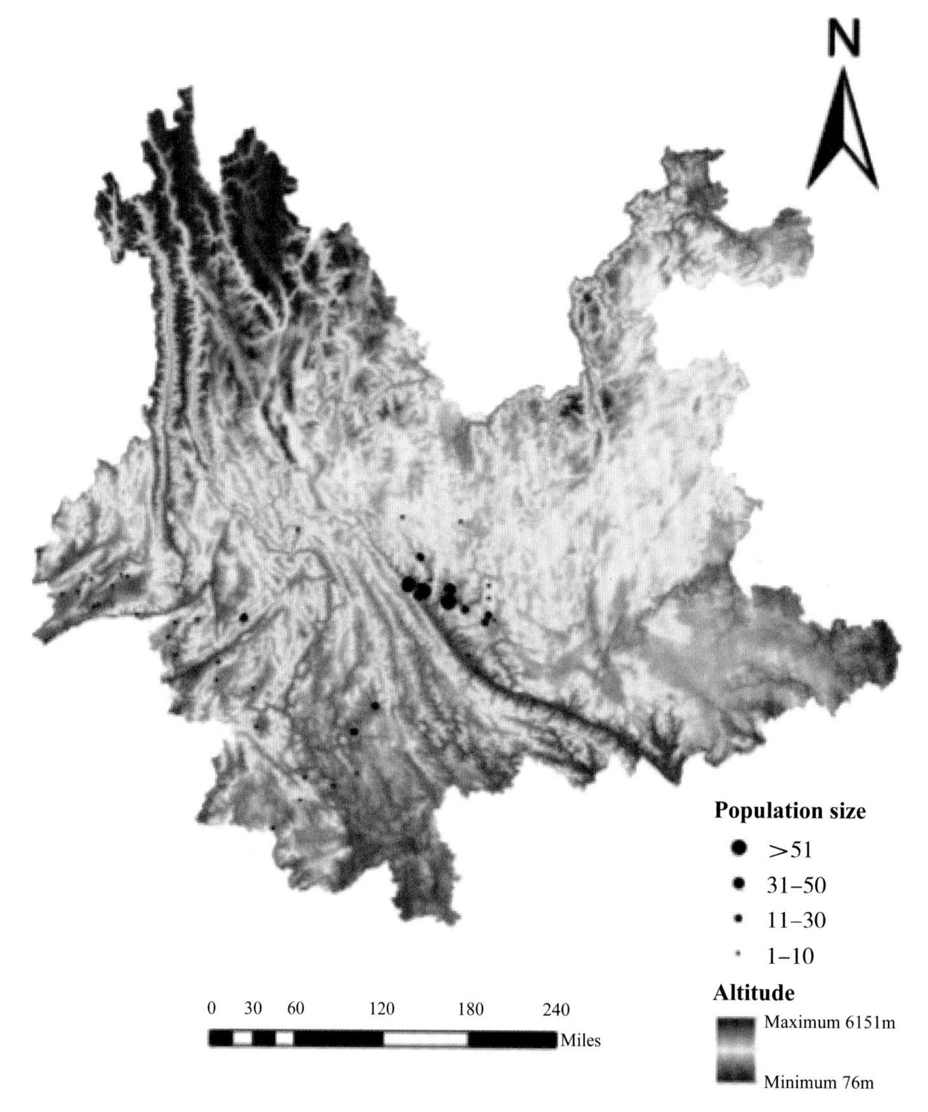

▲ Fig.1–4 Distribution and Population Size of Green Peafowl in Yunnan, China (Wu et al, 2019)
The map boundary in this figure doesn't constitute the basis for defining different areas

Yang Zhongxing et al. (2019) report that according to a nationwide green peafowl investigation implemented by the Kunming Animal Research Institute of the Chinese Academy of Sciences in 2014–2017, green peafowl are only distributed in 22 counties in 8 prefectures (cities) in central, western, and southern Yunnan in China. It is estimated that the number of the wild population is less than 500, which makes the green peafowl one of the most endangered wildlife species in China at present. According to the 2018 combined investigation of the Yunnan Forestry Department and the Kunming Animal Research Institute of Chinese Academy of Sciences, the population of the green peafowl in Yunnan Province is somewhere between 485–547, distributed over 19 counties in 6 prefectures and cities.

(2) Distribution and population of wild green peafowl in local areas

① Distribution and population of wild green peafowl in Gaoligong Mountain.From 1999–2005, the distribution of the green peafowl was recorded in an area 400–1250 m above sea level in the southern part of Gaoligong Mount by Ai Huaisen where the species were categorized as "rare" (Ai Huaisen, 2006).

② Distribution and population of green peafowl in the Konglonghe Prefectural Nature Reserve of Shuangbai, Chuxiong.From November 2014 to July 2015, Wen Yunyan and others carried out investigative and monitoring efforts focused on green peafowl in the concentrated distribution area in the Konglonghe Prefectural Nature Reserve of Shuangbai, Chuxiong, using the plotting method and infrared trigger automatic camera (20 infrared cameras were set up at 33 monitoring points). After analyzing the photos and videos that were obtained, it was concluded that there are 27 adult green peafowls and 29 chicks in the nature reserve (Wen Yunyan, 2016).

Fu Changjian et al. (2019) report that there are about 58 green peafowls gathered in 9 groups in the Konglonghe Prefectural Nature Reserve, making it the largest green peafowl habitat reserve in China before the Qinghua Nature Reserve in Weishan.

③ Distribution and population of green peafowl in the Qinghua Provincial Nature Reserve of Weishan, Yunnan Province.From October 2016 to September 2017, Li Binqiang and others installed 6,377 infrared cameras at 28 monitoring points in the core and buffer zones of the nature reserve in Qinghua, Weishan Yunnan Province. Among 1,692 usable distinct photos, 563 photos were of animals and 1,129 were of birds. However, the green peafowl itself, as the primary conservation animal of Qinghua Reserve, stands as an unsolved mystery in this survey. (Li Binqiang et al., 2018).

According to Fu Changjian et al. (2019), at present, there is only one protected area in China where the green peafowl are an animal under special protection. That is the Qinghua Provincial Nature Reserve of Weishan, Yunnan Province. The preserved area in Yunnan Province is located in the neighborhood of Beiyinqing, Huangjiafen and Baoziwo, Qinghua Village, 47 km from the Weishan County. Its geographical coordinates are 24°49′45″–25°10′0″N, 100°11′35″–100°14′50″E and was built in 1988 to become one of the provincial nature reserves in 1997. The conservation area covers a range of 100 hm^2. Its highest point is at an elevation of 2010.2 m and the lowest point can be found along the Yangjiang River, about 1146 m above sea level. A three-dimensional climate is prominent with an abundance of water, since the Longfeng and Zhongyao rivers flow through this reserve. About 20–30 green peafowls are safeguarded in the preserved area. Considering the small size of the conservation area and frequent human disturbance, it is a blessing that more protective measures are taken to guarantee a better survival rates of the green peafowl. However, related studies and reports on these protective measures along with the population numbers and density of green peafowl in these conservation zones have not been published.

④ Distribution and population of wild green peafowl along the Ruili-Menglian Expressway.Li Ruinian and Lin Haiyan (2018) made use of interview investigations, infrared cameras, and audio monitoring methods to focus on wild green peafowl along the Ruili-Menglian Expressway. Along the expressway, Longling and Zhenkang counties come into view. There are two green peafowl distribution points, which are located in the Jiangzhongshan District of the Xiaoheishan Provincial Nature Reserve of Longling and the Zhuwa District of Nanpenghe Provincial Nature Reserve of Zhenkang, both with less than 20 green peafowls.

⑤ Distribution and population of wild green peafowl in Xinping County, Yuxi City.Between January and December 2017, Wang Fang and others used infrared cameras to investigate and monitor wild green peafowl in Xinping County, Yunnan Province. A total of 96 infrared cameras were set up in a potential distribution area of wild green peafowl, which were operated over a total of 11482 working days, with images of wild green peafowl taken at 37 sites. 1370 distinct, effective photos of green peafowl were captured. According to the collected data, combined with 3S technology, a distribution map of green peafowl was made. Based on the results, wild green peafowl in Xinping County displayed a scattered shape across 6 districts of 5 towns/streets, such as in Guishan Street, Zhelong, Laochang, Xinhua, and Yangwu. These areas are all located outside of conservation areas, and the fragmentation of these habitats leads to the inability to communicate among various groups (Fig.1–5), making it more urgent than ever to carry out the work of protecting the green peafowl habitats (Wang Fang et al, 2018).

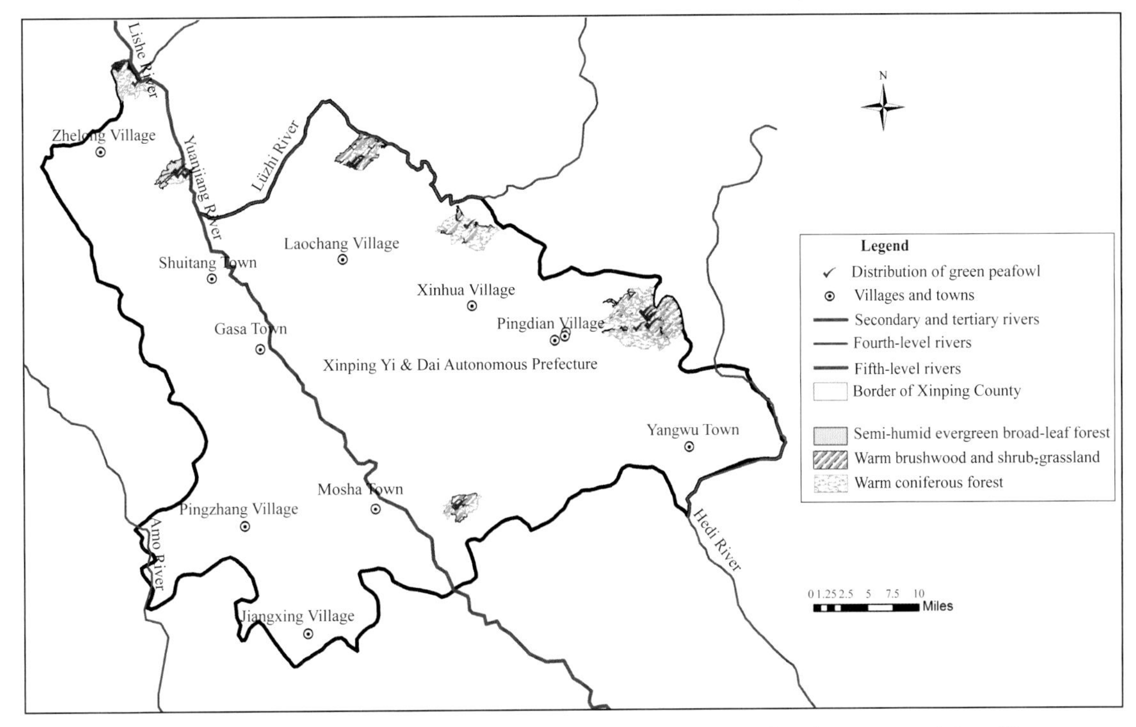

▲ Fig.1–5 Present Distribution of Wild Green Peafowl in Xinping County (Photo by courtesy of Chen Mingyong)
The map boundary in this figure doesn't constitute the basis for defining different areas

1.5 Investigation on Population and Distribution of Wild Green Peafowl in Xinping, Yuxi City, Yunnan Province

1.5.1 Investigation Methods

(1) Interview investigation

For green peafowl, as large, brightly colored, ground-dwelling birds, the local residents are very familiar with them. It is easy to identify the green peafowl for people in different age groups and with diverse cultural backgrounds. Anyone who's ever seen them has little chance to forget such an experience. Moreover, green peafowl distribution areas are places that are mainly inhabited by ethnic minorities. With honesty as their trademark, local residents can truthfully and accurately reflect various situations they've encountered in the past. Therefore, it proves necessary and efficient to investigate the distribution of the wild green peafowl by using interview methods.

Interview investigation steps:

① Form preparation and improvement: Firstly, the questionnaire form concerning the distribution of the wild green peafowl is compiled, and the form is modified and perfected after a pre-investigation to ensure that all fields in the form can reflect the distribution information of the green peafowl comprehensively, accurately, and concretely.

② County-level investigations: The leadership, technicians, and administrative staff of the Forestry and Grassland Bureau of Xinping County and the Ailaoshan National Nature Reserve in Yunnan Province are investigated to obtain the preliminary information on the distribution of the green peafowl.

③ Town-level investigations: According to the distribution information obtained in these county-level investigations, more detailed interviews and surveys are carried out among the leadership, management personnel, and patrol staff of the forestry stations and nature reserve management stations in the relevant towns to further pinpoint the villages where wild green peafowl are most likely to be so that more can be learned about corresponding village groups.

④ Village-level investigations: According to the previously obtained information, leaders of related village groups, forest rangers, hunters, graziers, and ordinary villagers who are familiar with wild green peafowl became respondents to a new round of interviews. The proper area, scope, and methodology of the field survey can be nailed down based on the provided distribution information.

⑤ Distribution information obtained from interview investigations is carefully checked to ensure that the information is true and useable. Based on visits made to the forestry staff at the county and township levels, visits and interviews are conducted to seven villages and towns (streets) initially identified as having or likely to have a distribution of the green peafowl, and village-level investigations are stopped in five towns (streets) where the distribution of the green peafowl was determined as zero. During January—June 2017, a total of 42 people from 23 village (residential) groups in 7 villages and towns (streets) in Xinping County have been interviewed. The interviews mainly record whether the interviewees have seen the green peafowl, the time, place, and number of the green peafowl they have seen, and the name, age, occupation, and other information of the interviewees. In order to obtain more accurate distribution information, the sample size of the investigation is scaled up, especially in places where the interviewee has ever seen a green peafowl or heard calls made by a green peafowl.

（2）Line-Transect method

The line-transect method is used to select a certain number of routes in the research area to carry out a sample survey and then to record the related information about the physical presence of animals and the traces they leave on the line transect. Based on information regarding the landform of the mountain and the distribution of vegetation obtained during the pre-investigation, the line transect is set at a length of 3–5km, with a width of 25 m at one side. The investigation spans 6 months, from January to June 2017, which means not only parts of the winter and summer and the whole of spring will be captured, but the entire reproductive period of the birds can be observed. Sunny mornings or evenings are the best time to implement such research. By using the line-transect method, we can observe and record green peafowl and the traces they leave behind (foot prints, feces, feathers, food marks, imprints after they lie on the ground, calls, etc.), and by making full use of hand-held GPS and digital cameras, the time, place, habitat, and vegetation types of green peafowl can be accurately recorded, which is advantageous to analyzing their living environment and range of activities.

Based on the results of the interview investigation, line transects are set up in areas where there may be a distribution of the green peafowl, so that a total of 10 green peafowl monitoring line transects appeared in five towns (streets) in Xinping County, including in Laochang Town, Zhelong Town, Xinhua Town, Yangwu Town and Guishan Street. Once a month, research efforts to record peafowl and their traces are carried out along the fixed line transects. A total of 6 line transect investigations have been conducted over the period of January to June in 2017, and 60 pieces of evidence pertaining to the line transect investigation are obtained.

①Population density calculation.Population density is obtained through absolute quantity research or by sampling surveys of the number of individuals in a particular study area. According to the results of the above quantity research, the density calculation method of the strip maximum counting method can be applied to calculate the population density of the green peafowl.

Strip counting method (Xu Long et al., 2003):

$$D=\frac{n}{2LW}$$

In the above formula, D stands for the density of bird population, n for the number of green peafowl on the line transect, L for the total length of the line transect, and W for the width of one side of the line transect.

② Population calculation. Population quantity can be calculated according to the formula $N=D\times A$ (Song Zhiyong et al., 2018) in which N stands for the quantity of the green peafowl, D for the population density, A for the distribution area of the green peafowl in the research area, and the deviation is ± 5%.

（3）Infrared camera method

① Infrared camera layout：Locations for the infrared cameras are determined based on the locations where interviewees have seen activity of green peafowl and where green peafowl and traces of their activity have been sighted on the line transect investigation. As a result, 96 automatic infrared cameras are installed in five villages and towns/streets of Xinping County in a 1 km × 1 km grid (Zhelong, Xinhua, Laochang, Yangwu and Guishan Street). The total coverage of the infrared cameras is 96 km^2. To adjust for different topographic conditions, infrared cameras are generally installed at the trunk, 40–100 cm away from the tree's roots, and a block of wood is cushioned between the camera and the trunk, so that the lens can maintain a predetermined angle with the ground from which to capture images. This way, the state of the green peafowl habitat at the installation position of the infrared camera is undisturbed, and the peafowl are unaffected (Wang Fang et al., 2018). There are two kinds of infrared camera models that were used in the investigation: the Lieke ERE-E1B and the Lieke Ltl–6210. The shooting mode of the infrared camera is set to "photo and video." For each independent event, three photos and one 10–second video is taken. The quality parameter for the photos is 12M, while the picture quality parameter for the videos is 1080P. According to different environmental and lighting conditions, the sensitivity of the infrared camera is set to either medium or low. The installation time of the infrared cameras ranges from January 2017 to June 2019, and the data from the infrared cameras installed in the field is retrieved in batches every month. Among all the data collected from the infrared cameras, data from January 2017 to December 2017 is relatively complete, so this period was selected for data analysis.

② Data processing method：Generally, the data retrieval of the infrared camera is finished once a month, and then all the data is stored into a large-capacity hard disk. When retrieving the data, research workers need to also attend to other tasks, including checking the infrared cameras to ensure they can continue to operate without issue, while also making sure to replace the camera battery and storage cards according to each camera's situation.

By using Bio-Photo, the image processing software designed for infrared cameras, the time information of extracting image data is automatically processed. The photos and videos of the green peafowl were named according to the unified format, and the species showcased in the data are artificially classified and identified.

One infrared camera installed in the field working normally for 24h is recorded as 1 working day, and the total effective camera working days is the sum of the working days of all the cameras. For the same infrared camera, an independent and effective event happens each time a different animal is captured on film. If the same animal is captured continuously, all images within 30 mins are recorded as only one independent and effective event (Yuan Jingxi et al., 2016).

③ Entity Identification Criteria: The photos/videos of the green peafowl taken by infrared cameras are identified via entity identification method. The green peafowl in different villages and towns and different regions of the same town or village are classified into different populations. The wild green peafowl in the same region are mainly distinguished by individual size, tail screen length, feather color, and time to appear at different infrared camera sites.

1.5.2 Results of the Investigation on the Population and Distribution of Wild Green Peafowl in Xinping County

(1) Results of the interview investigation

All 42 interviewees in the investigation came from 19 village (residential) groups in seven villages and towns (streets) in Xinping County. There were 34 interviewees from 15 village (residential) groups in five villages and towns (streets) who described the distribution of the green peafowl, while 8 interviewees from 4 village (residential) groups in Pingzhang Town and Mosha Town claimed that were no distribution of green peafowl (Table 1-1). Through the compilation of interview data, the number of green peafowl was estimated to be between 106–133, of which Laochang Town accounted for 10–20, Guishan Street 20–33, Yangwu Town 7, Zhelong Village 66, and Xinhua Village 3–7.

Table 1–1 Statistical data of green peafowl by interview investigation

Location	Number of village (residential) groups interviewed	Number of interviewees (person)	Results (the number of the green peafowl)
Laochang Village	3	11	10–20
Guishan Street	4	5	20–33
Pingzhang Town	3	6	0
Yangwu Town	4	6	7
Zhelong Village	3	9	About 66
Xinhua Village	1	3	3–7
Mosha Town	1	2	0
Total	19	42	106–133

(2) Investigation results based on the line-transect method

A total of 6 investigations were carried out over 10 line transects, and 13 green peafowl were recorded. These green peafowls were found in the Mashali Group, Zhuanmudu Hamlet, Laochang Village(2/event), Tadale, Hechamo Group, Zhuanmadu Hamlet, Laochang Village (1/event), Dajing Door, Daxiang Farm, Xiangyang Hamlet, Zhelong Village (1/event), Jianshanxiaojing, Zhongshan Group, Xiangyang Hamlet, Zhelong Village (3/event), Luohan Base, Maluzhai Village, Yangwu Town (1/event), Shuanglongqiao Group, Yani Community, Guishan Street (4/event) and Dahongjing, Dishebai Group, Daiwei Hamlet, Xinhua Village (1/event). In addition, 10 bird calls were heard, 20 footprints and 6 piles of feces were found, and 53 fallen feathers from peacocks shed after their breeding window were collected, and a breeding nest built by green peafowl in the field along with 3 peafowl eggs were also found (Table 1–2). The discovery of this evidence provides important guidance for the installation of infrared cameras.

Table 1–2 Statistical data investigated by line-transect method

Number	Position	Length of line transect (km)	Body	Feather	Foot print	calls	faeces	Peafowl eggs
1	Mashali, Zhuanmadu Helmet, Laochang Village	4.8	2	15	0	0	0	3
2	Tadale, Heichamo Group, Zhuanmadu Helmet, Laochang Village	4.2	1	3	0	4	0	0
3	Dajing Door, Daxiang Farm, Xiangyang Helmet, Zhelong Village	3.5	1	0	0	0	0	0
4	Jianshanxiaojing, Zhongshan Group, Xiangyang Helmet, Zhelong Village	3.8	3	7	0	0	0	0
5	Shaba, Xin Road, Dachunhe Group, Yao Helmet, Zhelong Village	4.6	0	0	9	3	2	0
6	Luohan Base, Maluzhai Village, Yangwu Town	4.2	1	4	0	0	0	0
7	Dhahe Mountain, Shuanglong Bridge, Guishan Street	3.5	1	11	0	0	0	0
8	Liangzi, Doujipo, Shuanglong Bridge, Guishan Street	3.4	2	0	11	0	4	0
9	Daoche Farm, Shuanglong Bridge, Guishan Street	3.9	1	3	0	0	0	0
10	Dahongjing, Dishebai Group, Daiwei Hamlet, Xinhua Village	4.5	1	10	0	3	0	0
	total	40.4	13	53	20	10	6	3

A total of 13 green peafowls were recorded in 6 investigations by using the line transect method. Based on the particularity of the bird investigation, it is difficult to estimate the relative density of birds by the trace method, so it is ultimately calculated based on the number of birds that were actually observed. According to the number of green peafowl, the relative density of the green peafowl on each line transect can be calculated, and their average population density over 10 line transects was found to be 0.664 per km^2.

(3) Investigation results based on the infrared camera method

From January to December 2017, all of the photographs and videos taken by the 96 infrared cameras were retrieved in eight stages, and the related image data was collected and organized. Results show that the infrared cameras worked for a total of 28,800 working days and took 195,167 photos and videos, of which 21,107 are of green peafowl, accounting for 10.8% of the total photos/videos. After rigorous photo/video identification, a total of 2,309 distinct and effective photos of 29 species of animals, such as green peafowl, jungle fowl, silver pheasants, and wild boar, and 1,378 distinct and effective photos of green peafowl were taken. After screening the image data of the green peafowl taken by the infrared cameras, the investigation results of the infrared cameras were obtained (Table 1–3), the photo and video of 126 green peafowl was taken.

Table 1–3 Numbers of Wild Green peafowls distributing in each area recorded by infrared cameras

Towns and Villages	Numbers of Wild Green Peafowls	Male	Female	Baby
Zhelong Village	82	19	57	6
Laochang Village	9	3	6	0
Xinhua Village	4	0	1	3
Guishan Street	26	10	16	0
Yangwu Town	5	2	3	0
Total	126	34	83	9

(4) Population of the green peafowl in Xinping County

According to the investigation, the number of green peafowl in the area ranges from 106–133, and photos/videos of 126 green peafowl were taken using the infrared camera investigation method. Based on population density (0.664 per km^2) and habitat area (145.11 km^2) of the green peafowl obtained from the line-transect method investigation, the number of the green peafowl was concluded to be between 91–101 (deviation of ± 5%) (See Table 1–4). Results from the infrared cameras are more accurate, so the number of the green peafowl in Xinping County has been determined to be 126.

Table 1–4 Statistical data of wild green peafowl in Xinping County

Towns and Villages	Infrared Camera Quantity (set)	Number of green peafowl based on interviews	Number of green peafowl taken by infrared cameras	Number of green peafowl based on line transects	Final number of green peafowl based on all-round analysis
Zhelong Village	47	66	82		
Laochang Village	11	10–20	9		
Xinhua Village	6	3–7	4		
Guishan Street	12	20–33	26		
Yangwu Town	20	7	5		
Total	96	106–133	126	91–101	126

1.5.3 Discussion on the Population and Distribution of Green Peafowl in Xinping County

(1) Population, distribution and protection of green peafowl in Xinping County

In this study, more comprehensive and detailed data is obtained on the population and distribution of green peafowl in Xinping County. The results reveal that: ① Green peafowl are mainly distributed over 6 areas of 5 villages and towns(streets) in Xinping County(Fig.1–5) and compared to documents recorded by Wen Xianji et al. in 1995, the number of the green peafowl present shows a clear increase, from 40–60 to 126 with 9 larvas included. ② The main distribution point of green peafowl in Xinping County is still Zhelong Village, the same place indicated in the results of previous investigations (Wen Xianji et al. 1995; Yang Xiaojun et al. 2017). ③ Guishan Street, Yangwu Town and Xinhua Village are the new distribution points of the green peafowl found in Xinping County. ④ The infrared camera investigation method and the line-transect investigation method are newly-added methods to obtain information about the current population of the green peafowl in Xinping County more comprehensively.

From the distribution map of the green peafowl, it can be seen that green peafowl are mainly scattered on the periphery of Xinping County, meaning green peafowl are located far away from each other, leading to an inability to communicate between the various populations. In this case, the green peafowl in Xinhua Village and Yangwu Town find themselves in a huge extinction risk. The finding that green peafowls in Zhelong Village and Laochang Village are distributed at the junction of Xinping County and Shuangbai County of Chuxiong City is in agreement with the investigation results concluded by Han et al., (2007).

The results of this investigation, when combined with previous data and literature in conjunction with the recent investigation data concerning the population and distribution of wild green peafowl both provincially and nationally as organized by the Yunnan Forestry and Grassland Bureau and Kunming Animal Institute of the Chinese Academy of Sciences, the wild green peafowl and its habitat in Xinping County are of great conservation value.

(2) Methods of investigation and research on population and distribution of wild green peafowl

In this investigation, the method of interview investigation, the line-transect method and infrared camera investigation method were all put into place. The advantages of using three investigation methods were that they made full use of different elements at different stages of the investigation to make the results more reliable. First, the interview investigation had access to the general activity area of green peafowl. Then, line-transect investigations were used to observe traces of peafowl activity and to provide important reference for setting up the infrared cameras. Finally, infrared cameras were set up to investigate. Information obtained in the interview investigation can not only provide clues for the installation of the infrared cameras and line transects, but also stands as a reference to the population (the population drawn from the interview phase is 106–133). Besides making the investigation more convenient, efficient, and accurate, infrared cameras are less disruptive for green peafowl and can take photos and videos directly while distinguishing them from other animals by image data comparison (population of the green peafowl at this time was 126). Application of the line-transect method and 3S technology provides more accurate data for determining the distribution and habitat range of green peafowl.

2

Morphological Characteristics of Wild Green Peafowl

2.1 Taxonomic Status of Green Peafowl

2.1.1 Taxonomic Status

(1) Species of Peafowl in the World

There are three species of peafowl throughout the world, namely, green peafowl (*Pavo muticus*) and blue peafowl (*Pavo cristatus*) which can both be found in Southeast Asia, and Congo peafowl (*Afropavo congensis*), which hail from the Congo Basin in Africa (Zheng Guangmei, 2002; Kong Dejun et al., 2017).

(2) Birds in the Pavo Genus

There are only two known species classified under the Pavo genus (*Pavo* Linnaeus, 1758), green peafowl and blue peafowl (Yang Lan et al., 1995).

(3) Taxonomic Status of Green Peafowl

Chinese people refer to green peafowl as *Kongque*, *Nuoyong* (in the Dai Language), *Yueniao*, or *Nanke* (in *Compendium of Materia Medica*) (Yang Lan et al., 1995; Kong Dejun et al., 2017). In the taxonomy of Aves, green peafowl belong to the order Galliformes, family *Phasianidae*, genus *Pavo* and species *Pavo muticus* (Linnaeus 1766).

(4) The Distribution and Population of Green Peafowl

Globally, the green peafowl's habitats stretch from northeast India to Yunnan China, through to Southeast Asia including Java. (Ma Jingneng et al., 2000). The global population of green peafowl dramatically declined in the 20th century. Nowadays, their population is scattered and many suffer from extirpation. Due to disturbances and changes in their habitats, green peafowl in Southeast Asia are still experiencing a declining population. It is estimated that there are 10000 to 19999 adult green peafowl worldwide (Kong Dejun et al., 2017).

2.1.2 Subspecies Classification

(1) Subspecies Classification of Green Peafowl

Three subspecies of green peafowl have been recorded thus far (Yang Lan et al., 1995; Kong Dejun et al., 2017; Hua Rong et al., 2018).

Java green peafowl (*P. m. muticus*) were once found in Java, Indonesia and Malaysia (Hua Rong et al., 2018). While they are still distributed throughout Java, it is believed that Java peafowl have become locally extinct in Malaysia and probably in Thailand as well (Kong Dejun et al., 2017).

Burmese green peafowl (*P. m. spicifer*) are native to Southeast Assam, India and west Myanmar (Hua Rong et al., 2018), although it is suspected that they have died out (Kong Dejun et al., 2017).

There is also *P. m. imperator*, whose Chinese name translates to the Indo-Chinese green peafowl (*P. m. imperator* Delacour 1949). This name, together with the Indo-Chinese green peafowl subspecies, is used by Hua Rong et al. (2018), but Zheng Zuoxin (2000) refers to *P. m. imperator* as the southern Yunnan green peafowl subspecies. This subspecies is found in east Myanmar, southwest China, Thailand and in Indo-Chinese regions. Kong Dejun et al. (2017) concluded that this subspecies has a geographic range covering south Myanmar, east Thailand, Cambodia, Laos, and Vietnam, before stretching north to the south of China.

(2) Subspecies of Green Peafowl in China

Most Chinese scholars agree that green peafowl, or *Pavo muticus*, are the only wild peafowl species in China (Yang Lan et al., 1995; Zheng Zuoxin, 2000; Zheng Guangmei, 2017; Kong Dejun et al., 2017) and *P. m. imperator* are the only peafowl subspecies in China (Yang Lan et al., 1995; Zheng Zuoxin, 2000; Zheng Guangmei, 2017; Kong Dejun et al.,

2017). However, some disagreement persists in their nomenclature in Chinese. While Zheng Zuoxin (2000) refers to *P. m. imperator* as the southern Yunnan green peafowl subspecies, Hua Rong et al. (2018) call them the Indo-Chinese green peafowl subspecies. Both terms are acceptable.

2.1.3 Conservation Status

Green peafowl were included in Annex I of *Convention on International Trade in Endangered Species of Wild Fauna and Flora* (CITES, 1995) as an endangered species (Zheng Guangmei, Wang Qishan, 1998). China listed green peafowl as a Class I National Key Protection Wildlife (Zheng Guangmei, Wang Qishan, 1998). Ma Jingneng et al. (2000) described them as "globally vulnerable" (Collaer et al., 1994) in A Field Guide to the Birds of China (Zhongguo Niaolei Yewai Shouce). In the China Wildlife Protection Practical Manual, Ma Jianzhang (2002) says that in the Convention on International Trade in Endangered Species of Wild Fauna and Flora that took effect on 19 July, 2000 after its 11th assembly, green peafowl are listed in Annex II. Yang Xiaojun et al. (2017) mention that the IUCN updated the status of green peafowl from vulnerable to endangered in 2009.

2.2 Morphological Characteristics of Green Peafowl

2.2.1 Physical Features

(1)Identification

Green peafowl are the biggest birds in the Phasianidae family. Green peacocks can grow 240 cm long. Adult peafowl can weigh up to 7700 g, and have a wingspan between 415 mm to 535 mm (Yang Lan et al., 1995).

Adult peacocks and peahens have similar features on their heads, and their plumage is mostly golden and emerald. Light blue skin can be seen around their eyes and their faces are bright yellow. Their bills are a dark brown while their lower mandible features a lighter shade. Their irises are reddish brown, and their tarsi and feet are brown.

Adult peacocks have crests on their crown, of which the middle is bright blue, the wide edge is emerald, and the base and shaft are brown. The neck, upper back and breast areas are copper while the base of the feathers is a dark purple with slight emerald edges. The lower back and abdomen are emerald and dotted by glistening violet spots. Peacocks have somewhere between 100 to 150 upper tail covering feathers that are each about a meter long. The top of the upper tail coverts are marked by bright blue and emerald eyespots. The tail feathers underneath are short and blackish brown. The primary covert and flight feathers

are a brownish yellow, with either a brown, blue-black, or emerald base (Fig.2–1, Fig.2–2). Peacocks have long tarsi. Peahens are smaller than peacocks at a length of 110 cm, and their body feathers are similar to those of peacocks except that they do not have trains. Their abdomens and backs are a dark brown tinged with a copper or a green metallic luster, while their backs are striped with dense and dark lines. The inner coverts and tail coverts look like the feathers on their backs, except that the color of the tail coverts is brighter. The tail feathers are dark brown with brown stripes and brown and white tips. A peahen's tail feathers are longer than their tail coverts (Yang Lan et al., 1995) (Fig.2–3, Fig.2–4).

Juvenile peacocks differ from adults in that their eye areas, chins, and throats are white(Fig.2–5, Fig.2–6).

▲ Fig.2–1 Profile of an adult green peacock (Photo by the green peafowl research team of Yunnan University using IR camera)

▲ Fig.2–2 Back view of an adult green peacock (Photo by the green peafowl research team of Yunnan University using IR camera)

▲ Fig.2–3 Profile of an adult green peacock in non–breeding period (Photo by the green peafowl research team of Yunnan University using IR camera)

▲ Fig.2-4 Back view of an adult green peahen (Photo by the green peafowl research team of Yunnan University using IR camera)

▲ Fig.2–5 Back view of a green peacock in post–breeding period (Photo by the green peafowl research team of Yunnan University using IR camera)

▲ Fig.2–6 Juvenile green peafowl (Photo by the green peafowl research team of Yunnan University using IR camera)

(2) Physical Features

① Head. Adult green peafowl have crests as long as 10 cm. These crests are bright blue in the middle and emerald green on the edge. The bases of the crest look like bluish green scales that sometimes show a violet luster. The exposed part around a green peafowl's eyes is a light cobalt blue, and the bare skin of their cheeks is bright yellow. Their bills are a blackish brown while their lower mandibles possess a softer shade (Yang Lan et al., 1995) (Fig.2–7).

▲ Fig.2–7 Close up of the head of an adult green peafowl Photo (Photo by the green peafowl research team of Yunnan University using IR camera)

② Neck. Adult green peacocks have golden copper feathers on their neck and front of their breasts. The base of these feathers is a dark violet blue. The edge of the vane is tinged with emerald green. The abdomen is darker with a dim violet luster (Fig.2–8).

▲ Fig.2–8 The feature color on the neck of a green peacock (Photo by the green peafowl research team of Yunnan University using IR camera)

③ Back. The backs of adult green peacocks look like pieces of green jade framed in black and are inlaid with oval aeneous spots. The wings are no less extravagant, and are covered with russet brown, blue-black and emerald green scale-like feathers. Feathers of different reflectivity catch and reflect sunlight to create iridescence (Fig.2–9).

▲ Fig.2–9 Feathers' colour on the back of a sub–adult green peacock (Photo by the green peafowl research team of Yunnan University using IR camera)

④ Wings and wing coverts. Adult green peacocks' primary coverts and primaries are brownish yellow with deep brown tips, while secondaries are deep brown and bear a bluish green metallic finish on the edge (Fig.2–10).

▲ Fig.2–10 Wings and wing coverts color of a green peafowl (Photo by the green peafowl research team of Yunnan University using IR camera)

⑤ The plumage of subadult peacocks during non-breeding seasons. Subadult peacocks are marked by their lack of long upper tail coverts, though the color of their tail is similar to that of adult peacocks. Generally, it takes male green peacocks three or more years to develop full trains. (Fig.2–11) Adult peacocks shed their upper tail feathers during non-breeding seasons. At the end of a breeding season, it is possible to spot short tail feathers with eyespots at the top of an adult peacock's back (Fig.2–12).

▲ Fig.2–11 A subadult green peacock that has not developed a train yet (Photo by the green peafowl research team of Yunnan University using IR camera)

▲ Fig.2–12 An adult green peacock that has shed its upper tail coverts by the end of the breeding season (Photo by the green peafowl research team of Yunnan University using IR camera)

⑥ Feature color on the backs of adult green peahens. The plumage of adult green peahens is darker than adult green peacocks. The feathers on the backs of peahens are a dark brown, with dense brown stripes and a bronze and green luster (Fig.2-13).

▲ Fig.2-13 The brown stripes on the back of an adult green peahen (Photo by the green peafowl research team of Yunnan University using IR camera)

⑦ Eyespots on the top of the trains of adult green peacocks. During the breeding season (usually from February to June), the ocelli—also known as eyespots—form patterns resembling "five-color" ancient Chinese coins on the peacock's train, marked by bright blue and emerald green. An ocellus consists of a violet oval frame, a yellow inner-ring, and an emerald colored fan-shaped marking at the center beneath a blue-black butterfly shape. The rest of the pattern is golden. Outside the oval frame are strands of varying lengths and colors including brown and purple.

▲ Fig.2–14 Eyespots on the top of upper tail coverts of an adult green peacock (Photo by the courtesy of Wang Fang)

⑧ Tail feathers and under tail coverts. The tail feathers of adult green peacocks are short and blackish brown. They are hidden under the train and can only be observed when the train fans out, during which the tail feathers also turn upward to support the train. The under-tail coverts are brown and look like down feathers. By contrast, adult green peahens have tail feathers longer than their tail coverts. The feathers are blackish brown with brown stripes and brown and white tips (Yang Lan et al., 1995) (Fig.2–15).

▲ Fig.2–15 Tail feathers revealed when a peacock displays (Photo by the green peafowl research team of Yunnan University using IR camera)

⑨ Upper tail coverts of adult green peacocks. During the breeding season (usually from February to June), green peacocks can possess as many as 100–150 extravagant upper tail coverts, each of which can be over one meter long. The top of the coverts are embellished with iridescent eyespots of bright blue and emerald （Fig.2–16）.

▲ Fig.2–16 Upper tail coverts of an adult green peacock during the breeding season (Photo by the green peafowl research team of Yunnan University using IR camera)

2.2.2 Trace Classification

（1）Physical Evidence

The best evidence to validate the distribution of wild green peafowl in field research is physical evidence such as any observed mature or juvenile peafowl and their eggs (Fig.2–17, Fig.2–18). Since wild green peafowl are alert creatures, the line-transect method usually fails to produce satisfying images, filming manually is not cost efficient enough and will cause interference to the green peafowl. Under such circumstances, IR cameras become an ideal option for researchers to capture peafowl in their natural environment.

▲ Fig.2–17　A peahen brooding (Photo by the green peafowl research team of Yunnan University using IR camera)

▲ Fig.2–18 Eggs of wild green peafowl discovered in a line transect (Photo by the courtesy of Wang Fang)

（2） Proxy Indicators

Proxy indicators can be used to estimate the presence of wild green peafowl. Indicators include footprints (Fig.2–19–Fig.2–21), feces (Fig.2–22), feeding marks (Fig.2–23), scratching marks or feathers. Spatial information about these traces should also be gathered. Useful data includes the time and place (longitude and latitude) of each discovery, distance to water sources, altitude, slope gradient and aspect, vegetation types, types of plants in the area, and whether green peafowl can feed on them. By running this data in an ecological model, researchers can assess whether the place is a potential habitat for wild green peafowl.

▲ Fig.2–19 Clear footprints of wild green peafowl on the beach (Photo by the courtesy of Chen Mingyong)

▲ Fig.2–20 Clear footprints of wild green peafowl on the beach (Photo by the courtesy of Chen Mingyong)

▲ Fig.2–21 Footprints and feathers of wild green peafowl on the beach (Photo by the courtesy of Wang Fang)

Green peafowl, together with jungle fowl, silver Phasianidae and domestic fowl are all members of the Galliformes order, Phasianidae family. Their footprints are similar, except those of green peafowl are larger due to their larger body size.

Like other birds, green peafowl have advanced a digestive system, which means there is usually no recognizable trace of plants or soil in their feces. It is also difficult to extract DNA from feces.

▲ Fig.2–22 Feces of wild green peafowl discovered in a line transect (Photo by the courtesy of Wang Fang)

▲ Fig.2–23 Feeding mark discovered in a line transect (Photo by the courtesy of Wang Fang)

(3) Feathers

Since adult peacocks molt covert feather from their tails at the end of the breeding season, discarded feathers can be found starting from the end of April. (Fig.2–24). Occasionally, researchers may find feathers from other parts of their bodies as well (Fig.2–25–Fig.2–28). Searching for feathers via line and pointing transects is necessary, since feathers serve as solid proof of wild green peafowl's presence in the area.

▲ Fig.2-24 Coverts of wild green peafowl found in a line transect (Photo by the courtesy of Wang Fang)

▲ Fig.2–25 An upper tail covert of wild green peafowl found on a line transect (Photo by the courtesy of Wang Fang)

▲ Fig.2–26 A fallen brown flight primary feather of wild green peafowl found on a line transect (Photo by the courtesy of Wang Fang)

▲ Fig.2–27 A fallen blackish blue flight coverts of wild green peafowl found on a line transect (Photo by the courtesy of Wang Fang)

▲ Fig.2–28 Down feather and coverts of wild green peafowl found on a line transect (Photo by the courtesy of Wang Fang)

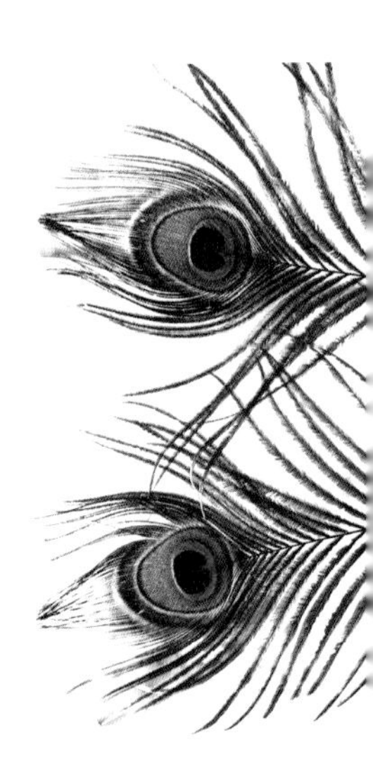

2.3 Green Peafowl and Blue Peafowl

2.3.1 Morphological Comparisons of Green and Blue Peafowl

Blue peafowl, also known as Indian peafowl, are the national bird of India. They are native to India, Sri Lanka and other countries and regions in South Asia. Blue peafowl are docile, highly domesticable and adaptable. Thus, it's a common practice to keep and breed blue peafowl (Nie Wangxing, 2017). Most peafowl kept by Chinese zoos are blue peafowl (Fig.2–29, Fig.2–30).

▲ Fig.2–29　A blue peacock (Photo by the courtesy of Wang Fang)

▲ Fig.2–30 The blue peahen (Photo by the courtesy of Wang Fang)

Blue peahens are smaller than peacocks. They are around 1 meter long and weigh 2.7 to 4 kilograms. Blue peahens do not possess trains. They have crests, chestnut brown crowns, blue-edged feathers on their napes, and white areas around their eyes, faces and throats. The lower parts of their necks, saddles and upper breasts are green while the rest of the upper body is olive brown (Fig.2–31).

▲ Fig.2–31 A hybrid peahen in Kunming Zoo (Photo by the courtesy of Wang Fang)

The crests of blue peafowl are fan-shaped. The eyes are striped with white at the top and bottom (Fig.2–32).

▲ Fig.2–32 The crest and tarsus of a hybrid peafowl (Photo by the courtesy of Wang Fang)

While green peafowl have green necks covered by scale-like feathers, blue peafowl have blue necks and filiform neck feathers (Fig.2–33).

▲ Fig.2–33 The filiform feathers on the neck of a hybrid peafowl (Photo by the courtesy of Wang Fang)

The green peafowl's crest feathers stand upright and in a cluster. The faces are light yellow. Peacocks have green scale-like feathers on their necks (Fig.2–34).

▲ Fig.2–34 Head and neck features of an adult green peacock (Photo by the green peafowl research team of Yunnan University using IR camera)

2.3.2 Other Color Patterns of Blue Peafowl

（1）White Peafowl

White peafowl are selectively bred from blue peafowl. They are a true breeding variety of blue peafowl. These albino peafowl cannot produce melanin due to abnormal tyrosinase, hence their light-colored irises and reddish eyes. White peafowl have a fan-shaped crest and pure white plumage (Fig.2–35).

▲ Fig.2–35 A white peafowl in Kunming Zoo (Photo by the courtesy of Chen Mingyong)

（2）Spalding Peafowl

Many color variations are made possible through the selective breeding. For instance, spalding peafowl can have white trains and eye areas, blue-black wing coverts and blue and white feathers on their bodies.

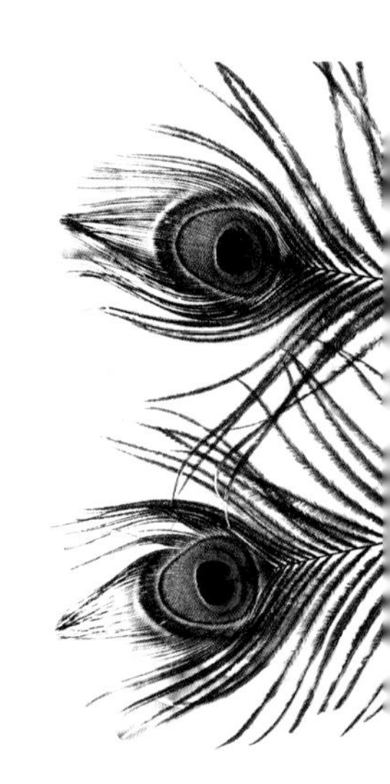

2.4 Present Situation of Peafowl in Captivity

In recent years, China has encouraged the integration of peafowl (green and blue) breeding and tourism. This is a promising industry for peafowl due to their high aesthetic, economic, and medical value. In addition, it is relatively easy to breed and raise peafowl, since they are adaptable creatures capable of living in colder or warmer climates. Therefore, the raising of peafowl is feasible in many parts of China. The ideal environments to raise peafowls are in villages with hilly, uncultivated surroundings. Many livestock farms have chosen to raise peafowl since they already possess the needed resources such as food and suitable land to do so. They also stand to mutually benefit from cooperation with the local tourism industry while catering to their agricultural businesses.

2.4.1 Present Situation of Green Peafowl in Captivity

Green peafowl have larger bodies and iridescent feathers. Famous for their appearance, green peafowl were once kept by many zoos around the world. But their population in the wild has dropped significantly and it is difficult to find wild green peafowl for breeding programs. In 1989, green peafowl were on the list of the first-class wild animals under special state protection in China. Unauthorized capture of green peafowl is illegal and comes with heavy penalties. According to our survey on Chinese institutes such as zoos and safari parks, there are few wild or bred green peafowl currently living in captivity, where inbreeding and hybridization problems with blue peafowl abound. More needs to be done in the field of pure breeding and the protection of their natural genetic state. Balancing conservation and cultivation, and conducting research on genetics and breeding will greatly benefit the protection of wild green peafowl and would come as welcome news to peafowl admirers everywhere. Such efforts are of great importance to this rare species.

We ran into a blue green hybrid peafowl in Manting Park, Jinghong City, Xishuangbanna Dai Autonomous Prefecture, Yunnan Province in 2012(Fig.2–36).

▲ Fig.2–36 The blue green hybrid peafowl in Manting Park (Photo by the courtesy of Chen Mingyong)

2.4.2 Present Situation of Blue Peafowl in Captivity

Blue peafowl are native to South Asia, Southeast Asia and the Indian subcontinent. Wild blue peafowl can now be found in Bangladesh, the Kingdom of Bhutan, India, Nepal, Pakistan, Sri Lanka, etc. They are the national bird of India and one of the national birds of Iran.

Blue peafowl are widely distributed in the coastal areas of Southeast Asia. People started breeding blue peafowl 3000 years ago for their aesthetic values (Dang Xinyan et al., 2014). Captive blue peafowl can be found in Australia, the Bahamas, New Zealand, the USA, and many other places around the world (Dang Xinyan et al., 2014). Yan Xiaojue (2009) deems that the captivity and breeding of blue peafowl dates back to 1987 in China. However, China's history of raising blue peafowl may be much longer in view of Wang Yanbo et al. (2018), due to the hypothesis that blue peafowl are in fact actually indigenous to Xinjiang. China has promoted the aviculture of blue peafowl and its relevant businesses. Although there are no statistics available, many businesses and producers in China are engaged in the raising of blue peafowl and the scale of their operation is expanding.

▲ Fig.2–37 A hybrid blue peacock in captivity for appreciation (Photo by the courtesy of Chen Mingyong)

Blue peafowl belong to the order *Galliformes*, family *Phasianidae*, and are one of the world's most attractive ornamental birds (Fig.2–37, Fig 2–38). Apart from their impressive appearance, the blue peafowl's appeal lies in other aspects as well. Its meat is edible, low in fat and cholesterol, and high in protein. A commercially bred blue peafowl is worth 1,000 RMB and a well-bred blue peafowl stands at 10000 RMB (Tang Songyuan et al., 2019).

It takes blue peafowl two years to be sexually mature, and peahen can lay 25 to 40 eggs, which weigh between 100–110g, and take around 28 days to hatch. Peahen look after the eggs during incubation and show a strong inclination to rear the chicks (Zhu Ziqiang, 1997).

In general, people raise blue peafowl for either their aesthetic or commercial value (meat and eggs). In many places, blue peafowl serve as an important component of the local arts for the eye-catching colors of their feathers and eggshells. In China, blue peafowl are not on the list of native species of China and can be raised, processed, and traded. There are established breeding methods. However, the National People's Congress and the Forestry and Grassland Bureau have banned the consumption of wild animals in the wake of the COVID–19 outbreak. Blue peafowl can still be raised in captivity, but only for aesthetic purposes.

▲ Fig.2– 38 A captive hybrid peacock displaying (Photo by the courtesy of Chen Mingyong)

3

Habitats of Wild Green Peafow in China

3.1 Habitat Characteristics

3.1.1 Topographical Features

Yang Lal.*et al.*, (1995) maintain that green peafowl live below an altitude of 2000 m in low mountains, hills, or river valleys. According to his finds in the Chuxiong Yi Autonomous Prefecture in central Yunnan, Xu Hui (1995) reported that green peafowl migrate to different elevations and slopes according to the changes of the seasons. For instance, they dwell on the sunny side of any given slope or in the river valleys during winter or spring. In the summer, green peafowl live mid-mountain (up to 1800 m above sea level) or deep in the bamboo groves there. Yang Xiaojul.*et al.* report that two green peafowl were spotted along the Zhibaluo River in Deqing County, Yunnan. One green peafowl was caught in Duosong village, Deqing County on 15 March, 1986. Zheng Guangmei and Wang Qishan (1998) believe that green peafowl naturally live in low hills and river valleys at around 1250 m above sea level. Luo Aidonl.*et al.* (1998) mapped the sites where wild green peafowl live in in Xishuangbanna, Yunnan, which include Dadugang, Jinghong City at 1300 m above sea level, and the Mangao conservation area, Menghai County at primarily 1300 m above sea level, with some at 1250 m above sea level and at flat side of the lower valley. Ma Jingnenl.*et al.* (2000) also reports that green peafowl can live at an elevation up to 1500 m. Ai Huaisen (2006) concludes that 2500 m above sea level and lower are suitable for the species belonging to the pheasant family and typical habitats for wild green peafowl lie in low altitude. Ai also has mentioned that wild green peafowl in the southern part of Gaoligong Mountain live between 400–1250 m above sea level while Kong Dejun and Yang Xiaojun (2017) contend that green peafowl inhabit low hills and river valleys below an altitude of 2500 m. Konl.*et al.* (2018) believe that wild green peafowl live at altitudes no higher than 2000 m and Fu Changjian (2019) claims that savannas or open plateaus with shrubs, broadleaf and coniferous trees are preferable for green peafowls.

On the basis of conclusions drawn from previous literature, we've found that Chinese wild green peafowl live at an altitude between 400 to 2500 m, though scholars' opinions vary from each other. At present, there is no monographic study on the vertical distribution of wild green peafowl in China. The investigation by the green peafowl research team of Yunnan University in Xinping Yi and Dai Autonomous County, Yuxi City, Yunnan Province revealed that the wild green peafowl are mainly distributed in low-middle mountain areas and river valleys below the altitude of 2500m (Fig.3–1), where the vegetation includes low-altitude sparse tropical rainforests along the lower slope, coniferous and broad-leaved mixed forests along the middle slope and coniferous forests (Yunnan pines) along the upper slope.

▲ Fig.3–1 The habitat of wild green peafowl in the river valley of the Shiyang River, upstream the Yuanjiang (Photo by the courtesy of Chen Mingyong)

3.1.2 Water Sources

Yang Xiaojul.*et al.* (2000) observed the habitats and behavior of green peafowl during spring in Jingdong County, Yunnan Province in 1996. The results show that among the 28 point transects, green peafowl appeared in a 100 m radius around water sources. Liu Zhal. *et al.* (2008) have similar findings in which green peafowl in the river valley of the Shiyang River had spring and autumn feeding sites close to water sources to go to after feeding. The feeding sites for green peafowl during spring are located near mountain streams or at the mouths of springs. Although water is plentiful in the river valley of Xiaojiang River, it was difficult to leave footprints because of the sandstone base; few traces of green peafowl were spotted there because the riverbanks were too steep for green peafowl to scatter after they drink. Li Xu.*et al.* (2016) have studied the habitat preference and spatial distribution profile of green peafowl during spring in the Konglong River conservation area in Chuxiong, Yunnan. Green peafowl in the conservation area usually feed on the sunny side of the slope in the valley, which features a milder gradient than the control sample area, and closer access to water sources and trails. Feces and footprints are frequently identified on trails and near water sources and support the above findings. Like other animals, peafowl tend to minimize energy consumption and maximize food resources when they forage. Thus, the topographic conditions mentioned are ideal as they provide access to water and save the peafowl's energy.

Based on these findings, it is clear that water resources are a key ecological factor for green peafowl. There is no known monographic study on how green peafowl consider local water sources when choosing habitats in their current habitat range.

The green peafowl research team of Yunnan University found clear footprints of green peafowl (Fig.3–2) near the mainstream of the Shiyang River, upstream the Yuanjiang on 7 April, 2017 during the teams' field research in Xinping County, Yuxi City, Yunnan Province. The footprints indicate frequent appearances of green peafowl in the area. The research also demonstrated that the riverbeds around the green peafowl's distribution area are their main water sources and the peafowl's courtship during the mating season (Fig.3–3).

▲ Fig.3-2 Footprints of wild green peafowl on the riverside of the mainstream of the Shiyang River, upstream the Yuanjiang (Photo by the courtesy of Chen Mingyong)

▲ Fig.3-3 Natural Sand bank alongside the Shiyang River where green peafowl form leks (Photo by the courtesy of Chen Mingyong)

3.1.3 Plants and Vegetation

(1) Research findings of wild green peafowl's preference for vegetation

Yang Lan et al. (1995) believe that green peafowl live in forests in tropical and subtropical regions, particularly in evergreen broad-leaf forests and deciduous broad-leaf forests in river valleys, and mixed coniferous forests and savannas. However, there is no trace of green peafowl in tropical rainforests. Luo Aidong et al. (1998) has found that most of the living wild green peafowl in Xinshuangbanna stayed in warm coniferous forests (*Pinus kesiya var. Langbianensis*), mixed coniferous forests, subtropical ever-green broadleaf forests, shrubwood and at the flatter ends of valleys that grew rice. Yang Xiaojun et al. (2000) has observed the spring habitat of green peafowl in Jingdong County, Yunnan Province and concluded that these areas mainly supported monsoon evergreen broadleaf forests, coniferous forests (*Pinus kesiya var. Langbianensis*), mixed coniferous forests, open thickets, scrub-grasslands, and farm fields. In March, April, October and November 2007, Liu Zhao et al. surveyed the foraging habitats of green peafowl upstream from the Shiyang River valley in Yuanjiang, Yunnan, using the line-transect and quadrat sampling methods. Altogether they measured 21 ecological factors which demonstrated that the quadrats in the peafowl's spring foraging habitat significantly differ from the control quadrats in terms of trail distance, tree type, and vine density, while the research done in autumn show insignificant differences. Comparison of ecological factors and logistic regression analysis showed that during spring and autumn, green peafowl always preferred foraging habitats with more fallen fruits, closer water sources, smoother gradients and larger tree coverage and diameter at breast height. Tree and grass coverage, distance to trails, residential areas of humans and forest edges are key ecological factors in the green peafowl's choice of foraging habitats during spring and autumn (Liu Zhao et al., 2008). Li Xu et al. (2016) has conducted research on the spring habitats of green peafowl in the Konglong River conservation area in Chuxiong, Yunnan and concluded that the foraging habitats there have higher tree coverage, canopy density, tree and vine diversity, and volume of fallen fruits. These factors show that green peafowl depend on the protection of tree coverage while foraging and prefer foraging habitats with abundant and diverse food sources, hence the diversity shown in trees and vines.

(2) The preferred vegetation type in Xinping County, Yuxi City, Yunnan Province

① Xinping Yi and the Dai Autonomous County (Xinping for short) is in the mid-southwest of Yunnan Province, on the east side of the middle section of Ailao Mountain between the latitudes of 23°38′15″–24°26′05″ N and 101°16′30″–102°16′50″ E. This county is the biggest one in Yuxi city, which covers an area of 4223 km^2. Xinping borders Eshan Yi Autonomous County to the east, Shiping County to the southeast, Yuanjiang County to the south, the Lvzhi River and the Chuxiong Yi Autonomous Prefecture Shuangbai County to the north. (Fig.3–4). Xinping has two sub-districts (Guishan and Gucheng), four towns (Yangwu, Mosha, Gasa and Shuitang), and six villages, (Xinhua, Laochang, Zhelong, Jiangxing, Pingdian and Pingzhang).

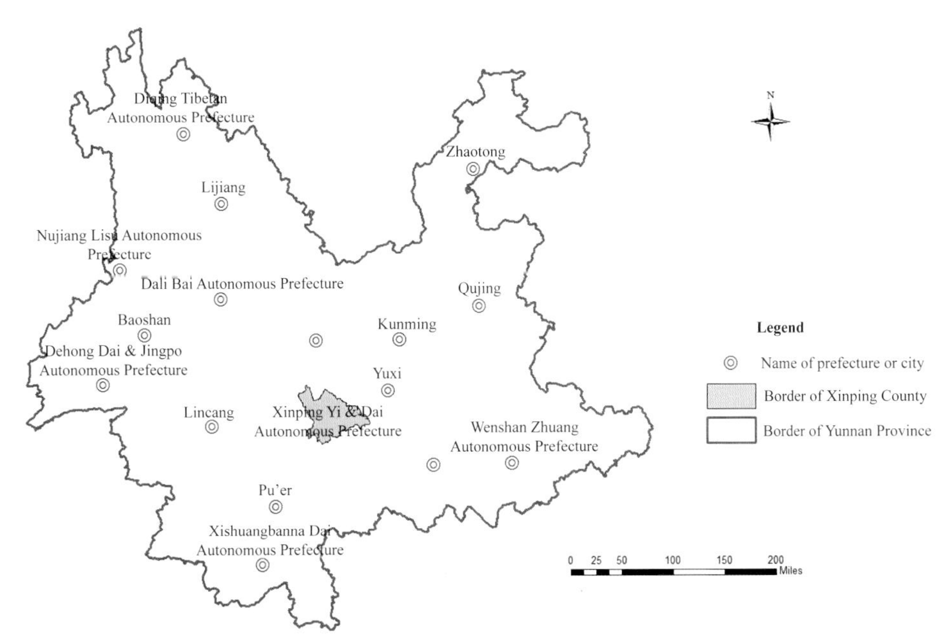

▲ Fig.3–4 The location of Xinping in Yunnan Province (Photo by the courtesy of Chen Mingyong)
The map boundary in this figure doesn' t constitute the basis for defining different areas

② Habitat Analysis. Yunnan University's green peafowl research team interpreted the 2015 satellite images of Yunnan Province and produced a distribution map of different types of vegetation in Xinping County. Based on interview and line-transect surveys, as well as IR camera data, the team has estimated that green peafowl are a constant presence in the region. Since vegetation type is used as vector data, and the altitudes as DEM (raster) data, the DEM

needs reclassification and areas below 2500 m need to be highlighted for the overlay. The raster image then needs to be converted into a vector image and superimposed over the map of types of vegetation in the wild green peafowl's habitats. Calculation and analysis done on ArcGIS 10.2 provided the types and areas of the green peafowl's habitats.

③Analysis Results. For the vegetation types included in this survey, only some of them showed traces of green peafowl, such as Sub-humid evergreen broad-leaf forest and the warm coniferous forests. We made a distribution map of green peafowl using 3S technology. It shows that wild green peafowl in Xinping County live in Xiangyang Hamlet, Zhelong Village; Yao Hamlet, Zhelong Village; Daiwei Hamlet, Xinhua Village; Zhuanmadu Hamlet, Laochang Village; Maluzhai Village, Yangwu Town; Shuanglong Hamlet, Guishan Street. The distribution of these habitats are scattered and disconnected (Fig.3–5, Fig.3–6).

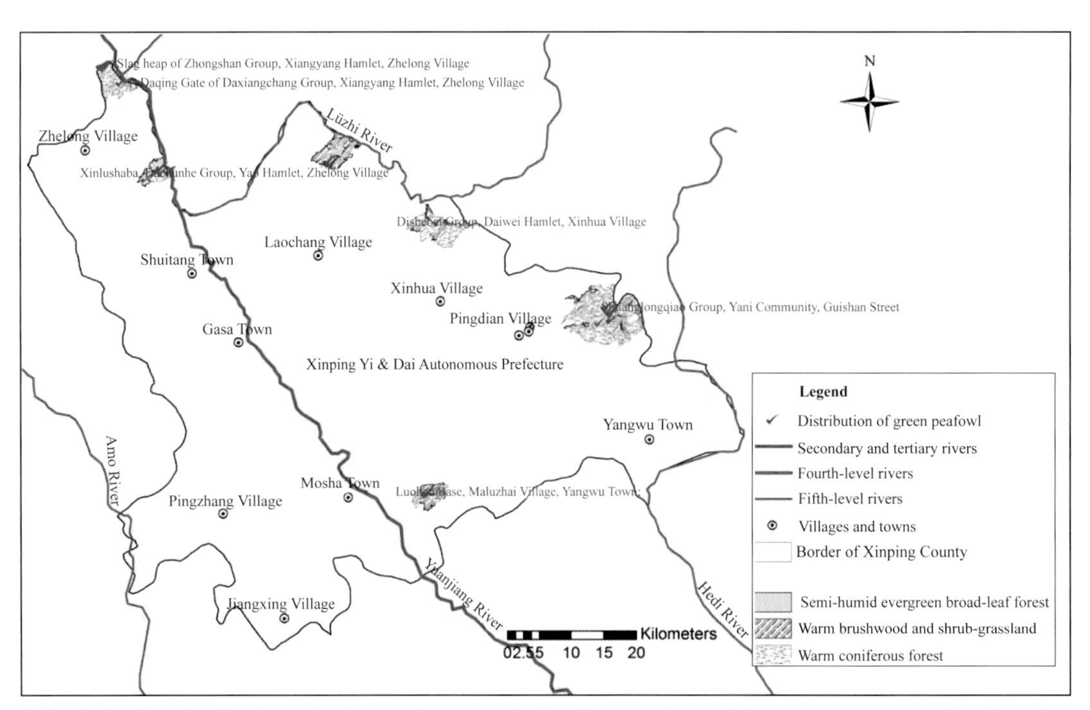

▲ Fig.3–5 The distribution and habitats of green peafowl in Xinping County (Photo by the courtesy of Li Zhengling)
The map boundary in this figure doesn' t constitute the basis for defining different areas

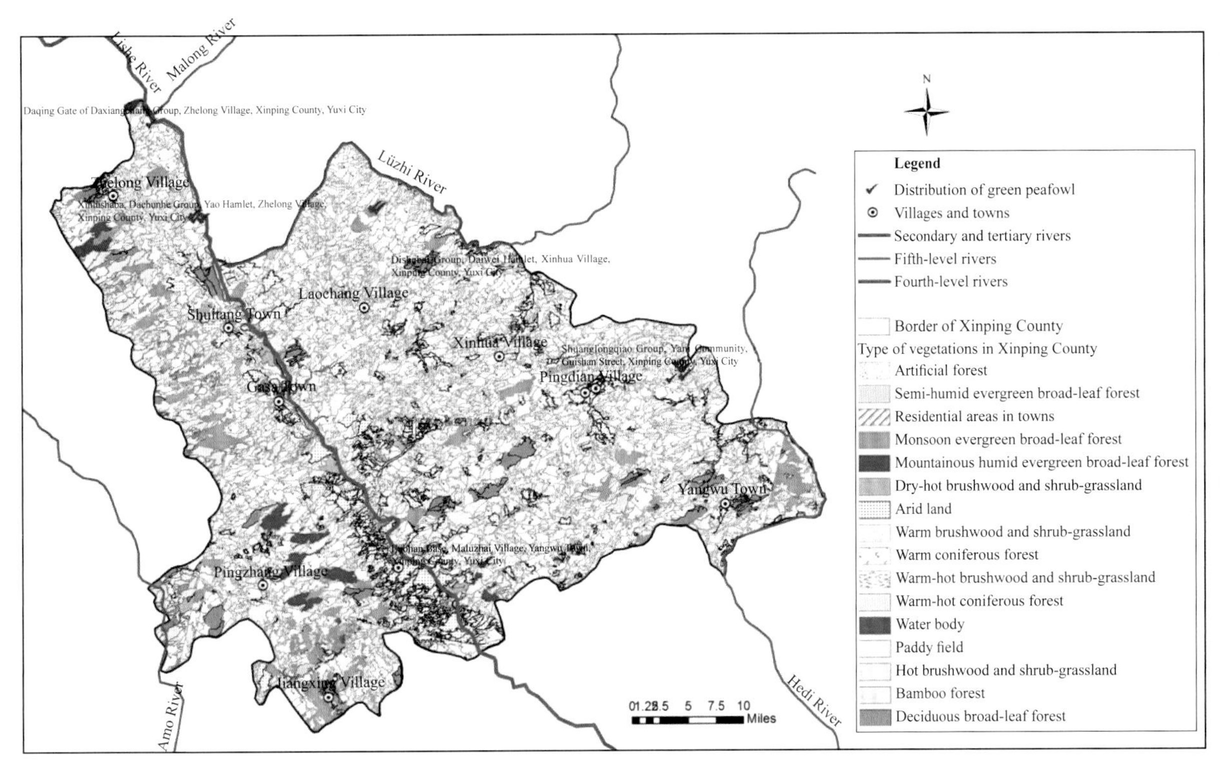

▲ Fig.3–6 The relative location of vegetation types and habitats in Xinping County (Photo by the courtesy of Li Zhengling) The map boundary in this figure doesn't constitute the basis for defining different areas

④ Habitat types and areas of wild green peafowls distributed in each area of Xinping County. After interpreting habitat types and areas in Xinping County, we conclude that the total area of the green peafowl's habitat in Xinping County is 14511 hm^2, accounting for 3.44% of the County (422300 hm^2). 67.25% (9759 hm^2) of the habitats are warm coniferous forest made up of Pinus yunnanensis. 28.54% (4141 hm^2) of the habitats are warm brushwood and scrub-grassland. 4.21 % (611 hm^2) of the habitats are sub-humid evergreen broad-leaf forest (Table 3–1).

Among the six habitats, Guishan Street is the largest with 6529 hm^2, accounting for 44.99% of the total habitat area. Guishan Street has two types of habitats—warm coniferous forest (5112 hm^2) and warm brushwood and scrub-grassland (1417 hm^2).

The habitat in Laochang Hamlet (1907 hm^2) is made up of warm coniferous forest (568 hm^2) and warm brushwood and scrub-grassland (1339 hm^2).

The habitat in Xinhua Hamlet is 2353 hm^2, consisting of 1712 hm^2 of warm coniferous forest and 641 hm^2 of warm brushwood and scrub-grassland. These two types of forest are interlaced, and green peafowl usually stay in the warm coniferous forest.

The habitat in Yangwu Town spans 1210 hm^2, and consists of 488 hm^2 of warm coniferous forest and 722 hm^2 of warm brushwood and scrub-grassland.

The habitat in Zhelong Hamlet is 2512 hm^2 and can be further categorized into Xiangyang Village and Yao Village. The habitat in Xiangyang Village has an area of 1489 hm^2 with 1467 hm^2 of warm coniferous forest and 22 hm^2 of warm brushwood and scrub-grassland. The habitat in Yao Village has an area of 1023 hm^2 with 412 hm^2 of warm coniferous forest and 611 hm^2 of sub-humid evergreen broad-leaf forest.

Table 3–1 Habitat types and areas of wild green peafowls distributing in each area of Xinping County

Habitats in each area in Xinping County	Habitat Types and Areas			
	Area (hm^2)	Warm coniferous forest (hm^2)	Warm brushwood and scrub-grassland (hm^2)	Sub-humid evergreen broad-leaf forest (hm^2)
Guishan Street	6529	5112	1417	0
Laochang Town	1907	568	1339	0
Xinhua Town	2353	1712	641	0
Yangwu Town	1210	488	722	0
Zhelong Town, Xiangyang Village	1489	1467	22	0
Zhelong Town, Yao Village	1023	412	0	611
Total	14511	9759	4141	611

⑤Preliminary analysis on the wild green peafowl's preference for vegetation. In Xinping County, wild green peafowl are mainly distributed in places below an altitude of 1000 m with mixed trees, shrub, coniferous and broad-leaf trees, within less than a 200 m vicinity of running water and dry soil. In particular, wild green peafowl are frequently found in forests of *Quercus acutissima* and *Cipadessa baccifera*. The wild green peafowl's favorite habitat is the monsoon rain forest, followed by the mixed coniferous forest, evergreen broad-leaf forest, and deciduous broad-leaf forest. By combining findings in the previous survey on habitat types done in Xinping County by the green peafowl research team of Yunnan University, it can be seen that the green peafowl's distribution range includes only warm coniferous forests, warm brushwood, and scrub-grasslands as well sub-humid evergreen broad-leaf forests. Moreover, they show a strong preference for warm coniferous forests.

⑥ The four types of spring habitats in Xinping County.

A. Evergreen broad-leaf forests where *Cipadessa baccifera* is abundant (Fig.3–7, Fig.3–8).

▲ Fig.3–7 Evergreen broad leaf forest (Photo by the courtesy of Wang Fang)

▲ Fig.3–8 Green peafowl habitat: open thickets (Photo by the courtesy of Wang Fang)

B. Monsoon rainforest that feature shrubs and small trees usually lying on the edge of forests. The forests are around 5 to 6 m tall. The trees and plant life include *Ficus microcarpa*, *Cipadessa baccifera* and *Quercus acutissima*; the shrubs include *Phyllanthus emblica*, *Portulaca grandiflora*, *Zanthoxylum nitidum*, *Pinus yunnanensis*, *Eupatorium odoratum*, *Eupatorium adenophora Spreng*, *Quercus acutissima*, *Eriolaena spectabilis*, *Madhuca pasquieri)* , *Ficus curtipes* and *Rauvolfia verticillate* (Fig.3–9).

▲ Fig.3–9 Green peafowl habitat: Monsoon forest (Photo by the courtesy of Wang Fang)

C. Mixed coniferous forest with plants like *Pinus yunnanensis*, *Mangroves*, and *Schima superba* which form a mixed pine and oak forest. Such habitats are usually found on hills. Apart from *Pinus yunnanensis*, there are *Quercus acutissima*, *Castanopsis fleuryi*, *Phyllanthus emblica*, *Ageratina, Chromolaene odorata*, *Aristolochia debilis*, *Portulaca oleracea* and *Geranium wilfordii* (Fig.3–10, Fig.3–11).

▲ Fig.3–10 Green peafowl habitat: mixed coniferous forest (Photo by the courtesy of Wang Fang)

▲ Fig.3–11 A green peafowl in a mixed coniferous forest (Photo by the green peafowl research team of Yunnan University using IR camera)

D. Deciduous broad-leaf forests with dominant plants like *Cipadessa baccifera, Buxus sinica, Chromolaena odorat and Stereospermum colais* etc. (Fig.3–12–Fig.3–15).

▲ Fig.3–12 A green peafowl in a deciduous broad–leaf forest (Photo by the green peafowl research team of Yunnan University using IR camera)

▲ Fig.3–13 Green peafowl habitat: forest of *Pinus Yunnanensis* (Photo by the courtesy of Wang Fang)

▲ Fig.3–14 A green peafowl in a forest of *Pinus Yunnanensis* (Photo by the green peafowl research team of Yunnan University using IR camera)

▲ Fig.3–15 Green peafowl habitat: underbrush (Photo by the courtesy of Chen Mingyong)

Adult green peafowls also live in various habitats, sometimes on roads (Fig.3–16), open Spaces at forest gaps (Fig.3–17), forest edges (Fig.3–18), brush forests (Fig.3–19) and under trees (Fig.3–20), but on the premise that these areas are safe and free from human interference.

▲ Fig.3–16 A green peafowl in the brushwood by the road (Photo by the courtesy of Chen Mingyong)

▲ Fig.3–17 Green peafowl in a coniferous forest (Photo by the green peafowl research team of Yunnan University using IR camera)

▲ Fig.3–18 Green peafowl on the edge of a forest (Photo by the green peafowl research team of Yunnan University using IR camera)

▲ Fig.3–19 Green peafowl in a broad–leaf forest (Photo by the green peafowl research team of Yunnan University using IR camera)

▲ Fig.3-20　Green peafowl under a megaphanerophyte (Photo by the green peafowl research team of Yunnan University using IR camera)

3.2 Feeding Habits

Zheng Zuoxin et al.（1978）believe that green peafowl feed mainly on birchleaf pears, fruits of *Rubus obcordatus*, rice, seedling and grass seeds, and also on insects such as crickets, grasshoppers, moths, and frogs as well as lizards. Lu Taichun et al. (1991) dissected a peahen found in Jingdongling Street in the city of Pu'er, in which they found mushrooms, young leaves, grass, termites, stink bugs, etc, but mostly mushrooms in the 102g of undigested food (Yang Lan et al., 1995).

The survey of the green peafowl research team of Yunnan University in Xinping Yi Autonomous County, Yuxi City, Yunnan Province reveals that the local peafowl have a varied diet consisting of flora and fauna, mainly succulent young leaves, berries, and insects, more specifically, plants and crops such as *Pyrus pashia*, fruits of *Rubus obcordatus*, leaves of saplings, buds, mushrooms, grass seeds, *Zea mays*, *Pisum sativum and Olyza sativa*. There were also insects such as grasshoppers, crickets, moths, termites, stink bugs, earthworms, and small animals like lizards and frogs. During spring, fruits of *Cipadessa cinerascens* (Fig.3–21–Fig.3–30), *Ficus curtipes,* and *Tectona grandis* become their primary sources of plant food, with the *Tectona grandis* being the peafowl's favorite. According to local forest ranger, Qi Guofang, the green peafowl there also eat the seeds of Yunnan olives (*Phyllanthus emblica*) (Fig.3–27, Fig.3–28). Wild green peafowl often appear in groups to forage in crop fields and show a preference for peas and yam. To help digestion, green peafowl will swallow grains of sand.

The feeding habits of wild green peafowl in China have been scarcely discussed based on our literature review. In addition, there is no published catalogue of the plant foods wild green peafowl in China feed on or prefer. To have a better understanding of the situation and habitat quality of wild green peafowl in China, more research needs to be done on wild green peafowl's feeding habits.

▲ Fig.3–21 Green peafowl's diet: *Cipadessa baccifera* (Photo by the courtesy of Wang Fang)

▲ Fig.3–22 Green peafowl's diet: *Ficus racemosa* (Photo by the courtesy of Wang Fang)

▲ Fig.3–23 Green peafowl's diet: *Ilex rotunda* (Photo by the courtesy of Wang Fang)

▲ Fig.3–24 Green peafowl's diet: *Woodfordia fruticosa* (Photo by the courtesy of Chen Mingyong)

▲ Fig.3–25 Green peafowl's diet: *Solanum torvum* (Photo by the courtesy of Wang Fang)

▲ Fig.3–26 Green peafowl's diet: *Opuntia dilleniid* (Photo by the courtesy of Wang Fang)

▲ Fig.3–27 Green peafowl's diet: *Phyllanthus emblica* (Photo by the courtesy of Wang Fang)

▲ Fig.3–28 Green peafowl's diet: *Phyllanthus emblica* (Photo by the courtesy of Wang Fang)

▲ Fig.3–29 Green peafowl's diet: *Woodfordia fruticosa* (Photo by the courtesy of Wang Fang)

▲ Fig.3–30 Green peafowl's diet: *Tectona grandis* (Photo by the courtesy of Wang Fang)

4

Ecological Habits of Wild Green Peafowl in China

4.1 Habitat Selection

4.1.1 Selection of Foraging Habitats

Research on the Habitat Selection and Spatial Distribution of Green Peafowl (Pavo muticus) in the Konglonghe Natural Reserve of Chuxiong, Yunnan Province by Li Xu, etc. (2016) illustrates that decisions made by green peafowl in regards to their foraging areas and night-roosting sites correlates with their activities and behavior. In the morning, green peafowl will usually fly down or glide to mountain valleys of lower elevations for drinking and foraging, then fly up to higher places for roosting in the evening. Green peafowl clearly share similar activity patterns with other pheasants in terms of their daily vertical movements. Unless the green peafowl is disturbed, it will always forage or night-roost within the same area. Food and concealment prove to be the biggest factors when considering the green peafowl's roosting and feeding habits. In particular, they prefer sunny slopes with gentler inclinations along mountain valleys and areas which are close to water sources and small paths since the canopy density—consisting of various kinds of tall trees and vines—will be higher there. Due to the fact that the coverage and diameter at breast height (DBH) of trees fundamentally affects the green peafowl's night-roosting site selection, they prefer forests where trees are closed and tall. Habitat selection of green peafowl interacts with environmental factors in various ways, thus when preferred foraging locations and ideal roosting sites differ in terms of their environmental characteristics, these sorts of daily vertical movements will start to emerge. Normally, green peafowl cluster in sites at higher elevation and gather in areas that offer ideal concealment conditions and abundant food and water resources (Fig.4–1–Fig.4–6).

▲ Fig.4–1 A green peacock is foraging in a sparse forest (Photo by the green peafowl research team of Yunnan University using IR camera)

▲ Fig.4–2 A green peahen is foraging along a mountain valley (Photo by the green peafowl research team of Yunnan University using IR camera)

▲ Fig.4-3　A sub-adult green peacock is foraging at the edge of a forest (Photo by the green peafowl research team of Yunnan University using IR camera)

▲ Fig.4-4　A green peacock is foraging in sand (Photo by the green peafowl research team of Yunnan University using IR camera)

▲ Fig.4–5 A green peahen frequently found foraging near the night–roost site (Photo by the green peafowl research team of Yunnan University using IR camera)

▲ Fig.4–6 A green peahen is resting in a night–roosting tree after foraging (Photo by the green peafowl research team of Yunnan University using IR camera)

4.1.2 Selection of Drinking Sites

The investigation by the green peafowl research team of Yunnan University in the Zhelong Township of Xinping Yi and Dai Autonomous County in Yuxi reveals that green peafowl's activities can usually be observed around rivers, streams, sandy areas, and ponds. Such activity patterns show that green peafowl are greatly dependent on water for their daily activities. In spring when rain is sparse, such behavior is especially obvious because green peafowl move up along valleys to forage in the day and find water sources to use at midday or in the afternoon (Fig.4–7).

▲ Fig.4–7 A stream near a site of green peafowl activity (Photo by the courtesy of Chen Mingyong)

4.1.3 Selection of Courtship Sites (Fig.4–8–Fig.4–9)

Each year in spring, green peafowl often choose open areas to cluster and adult males will unfold their tail screens to show off and court adult females. The main choices of such courtship sites are open areas in the forest (Fig.4–8), riverbank beaches (Fig.4–9), and others.

▲ Fig.4–8 A green peacock uses an open area in the forest to court a peahen in breeding season (Photo by the green peafowl research team of Yunnan University using IR camera)

▲ Fig.4–9 Green peafowl also select open sand beaches along riverbanks where they can avoid being interfered with as courtship sites (Photo by the courtesy of Chen Mingyong)

4.1.4 Selection of Night-roosting Sites

Research by Li Xu, etc. (2016) illustrates that green peafowl in the Konglonghe Natural Reserve of Chuxiong, Yunnan Province, select sunny slopes in higher elevations which are farther from water sources as their night-roost sites in the spring. This is because there is a limited variety of tall, closed trees, and they are lower in density. Additionally, there are fewer bushes, with fallen leaves and herbs constituting a larger proportion of available coverage, and vines and seeds are low in density. Members of the green peafowl research team of Yunnan University have found night-roosting sites by monitoring green peafowl with IR cameras and have also captured the image of a green peafowl resting in a tree (Fig.4–10).

▲ Fig.4–10 A green peacock is resting in a tree at night–roost sites (Photo by the green peafowl research team of Yunnan University using IR camera)

4.2 Activity Rhythm of Green Peafowl

4.2.1 Current Research on the Daily Activity Rhythm of Green Peafowl in China

There has been research and reports on the daily activity patterns of green peafowl in China. For example, Yang Xiaojun, etc., preliminarily observed the habitats, activities, and behaviors of green peafowl during the springtime in Jingdong County of Yunnan Province in 1996. By making observations using the line transect method, researchers recorded what they saw as well as any chirping sounds that they heard from 7:00–20:00 every day, and found the movements of 23 distinct green peafowl could be observed between 8:00–13:00 and 16:00–20:00, with their major feeding times ranging from 7:00–12:00 and 18:00–20:00, based on the activity frequency of green peafowl. There were two peaks of green peafowl activity in the morning and afternoon. Moreover, feeding accounted for 51.82% of all their activity, which is similar to those of captive green peafowl (Yang Xiaojun, etc., 1996). Accordingly, researchers believe that the similar activity rhythm between wild and captive green peafowl may possibly mean that the daily activity rhythms of various animals must have been formed over a long evolutionary process (Yang Xiaojun, etc., 2000). Li Xu, etc. (2016) adopted the line transect method and quadrat sampling method in the investigation of the Konglonghe Natural Reserve of Chuxiong, Yunnan Province in spring and found that green peafowl tended to fly or glide down from their night-roosts to mountain valleys at lower elevation to drink and forage in the morning (6:00 a.m. to 7:00 a.m.), before flying up to their night-roosting sites in higher elevation in the evening (5:30 p.m. to 7:30 p.m.).

4.2.2 Analysis on the Daily and Annual Activity rhythms of Wild Green Peafowl in Xinping County with IR cameras

Through the entirety of 2017, the Yunnan University's Green peafowl research team has analyzed the daily activity rhythm of wild green peafowl in Xinping County, Yuxi, Yunnan Province with IR cameras, and the findings are as follows:

(1) Research method

Analyze the activity rhythm of green peafowl according to images by IR cameras:

① Analysis of daily activity rhythm. A day is divided into 24 sampling periods at 1–hour intervals. Researchers count the independent probe number of each period and calculate the relative activity index (RAI) of each period.

$$RAI = \frac{Mg}{M} \times 100$$

Relative Activity Index (RAI):

In which *Mg* represents the independent image number of green peafowl in *g* period, and *M* represents the total independent images of green peafowl.

② Analysis of annual activity rhythm. The probability of animals shot by IR cameras is positively correlated to their activity indexes (Li Mingfu, etc. 2011), while RAI reflects the activity index of animals. Independent probe number is counted by month to compare RAI indexes of each month so that the annual activity rhythm may be analyzed.

(2) Research Results

① Daily activity rhythm. Activity intensity in each period every day is as following. Analyze 1,378 independent and effective photos of green peafowl in 2017, and calculate the activity intensity of green peafowl for each period (for instance, 6:00 a.m. to 6:59 a.m. was recorded as 6) (Table 4–1).

Table 4–1 Daily rhythm of green peafowl

Period	Number of Independent and Effective Photos	Relative Activity Intensity
6	63	4.57%
7	163	11.83%
8	149	10.81%
9	119	8.64%
10	77	5.59%
11	67	4.86%
12	55	3.99%
13	43	3.12%
14	50	3.63%
15	50	3.63%
16	75	5.44%
17	168	12.19%
18	209	15.17%
19	84	6.10%
20	6	0.44%
Total	1378	100%

The plotted daily activity rhythm shows that green peafowl move around from 6:00–21:00 with significant peaks and valleys. The peaks occur from 7:00–9:00 and 17:00–18:00, presenting a twin-peaks model in which the maximum relative activity intensity occurs between 7:00–18:00. The valleys tend to appear between 12:00–15:00, during which green peafowls tend to rest in shaded areas such as in or beneath trees. Green peafowl start to fly down from trees to forage at 6:00 every day and to rest in these trees at around 20:00 (Fig.4–11).

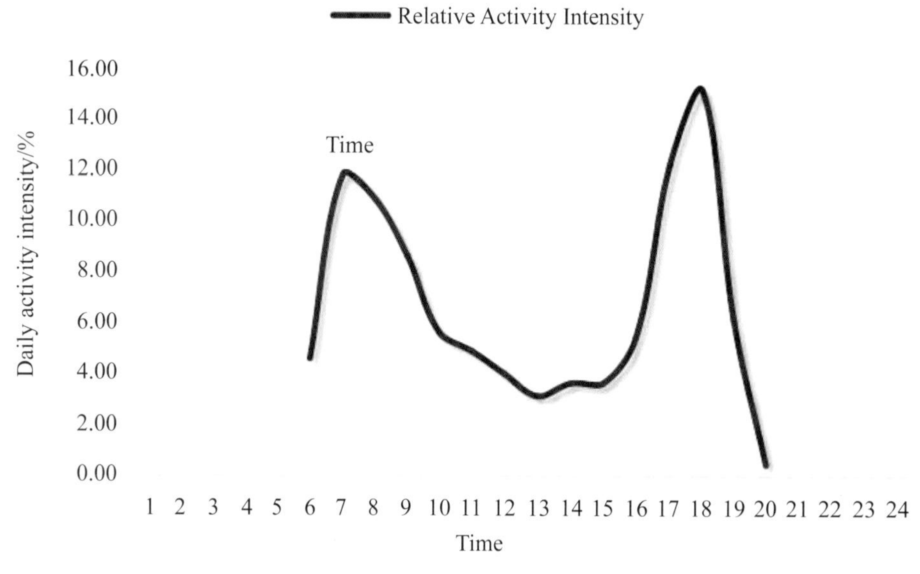

▲ Fig.4–11 Daily activity rhythm of green peafowl

② Discussions on daily activity rhythm: Over the entire year's analysis on IR camera data, we conclude that the movements of wild green peafowl in Xinping County can be observed primarily between 6:00–21:00, consistent with that of captive ones (Yang Xiaojun and Yang Lan, 1996) and to that of many pheasants (Zhao Yuze, etc., 2013; Zhou Xiaoyu, etc., 2008), while being slightly longer than that of wild green peafowl in Jingdong region, Yunnan Province which are active between 8:00–20:00 (Yang Xiaojun, etc., 2000). The main reason for this disparity can probably be traced to the differences between the technical means and methods adopted to document the peafowl, since IR camera monitoring is more accurate than just line transect surveying and monitoring. IR cameras monitor activities of green peafowl for 24 hours a day, which is especially useful at dawn and dusk when it becomes difficult for human beings to carry out surveys and conduct monitoring due to poor visibility. Furthermore, IR cameras monitor green peafowl statically, and thus avoid interfering with the findings. For this reason, IR cameras are particularly useful and are worth promoting for their applications in recording the daily activity rhythm of large birds, especially large pheasants.

According to research on the behavioral ecology of green peafowl on Java Island, green peafowl there are active from 5:00 to 18:00, which shows that the timetable for their active and resting periods is earlier than that of domestic ones. Researchers infer that, on one hand, green peafowl on Java Island and those in China are different sub-species which possess their own distinct living habits, while on the other hand, they are also affected by the different living environments and different timeframes for sunrise and sunset. Green peafowl on Java Island have a 4–5 hour midday rest period which occurs after their eating and drinking period in the morning and before their eating period in the afternoon. Such patterns are similar to those uncovered by domestic research because it is a mechanism for birds to elude direct sunlight and to rest.

A. 7:00–12:00 Foraging

Green peafowl move around frequently at dawn and dusk. At 5 and 6 in the morning, they leave the tall trees where they night-roost and glide to lower places or riversides for drinking or foraging. Wild green peafowl tend to chirp when they wake up to announce the start of a new day (Fig.4–12, Fig.4–13).

▲ Fig.4–12 A green peahen is foraging in the morning (Photo by the green peafowl research team of Yunnan University using IR camera)

▲ Fig.4–13 A green peacock is foraging in the morning (Photo by the green peafowl research team of Yunnan University using IR camera)

B. 12:00–16:00 Resting

When it is sunny and hot at noon, green peafowl tend to hide below thick bushes and dig sandpits: bathing themselves in sand—throwing sand onto their bodies for disinfection and sterilization, or lying there to preen themselves. Ancient people discovered such behavior early on: "They like to lie in the sand to bath themselves and enjoy comfort"(Fig.4–14–Fig.4–15).

▲ Fig.4–14 A green peahen is resting in the afternoon (Photo by the green peafowl research team of Yunnan University using IR camera)

▲ Fig.4–15 A green peacock is resting in the afternoon (Photo by the green peafowl research team of Yunnan University using IR camera)

After a short break at noon, green peafowl start to go out foraging until 20:00.

C. 16:00–20:00 Foraging

▲ Fig.4–16 A green peahen is foraging before the sunset (Photo by the green peafowl research team of Yunnan University using IR camera)

▲ Fig.4–17 A green peacock is foraging in the afternoon (Photo by the green peafowl research team of Yunnan University using IR camera)

Green peafowl remain alert even at rest, and during mating season, they will chirp loudly "ga-wo, ga-wo". The sound of their calls rings out continuously throughout the entire valley. People who are unfamiliar with such sound may certainly be frightened and misunderstand it as the sound of beasts or birds of prey. As evening approaches, green peafowl fly back in small groups to their night-roost sites and climb into tall trees for resting. In fact, green peafowl select their night-roost trees very carefully. Not only do they select comfortable and concealed trees, but also tall and canopied ones, which go hand in hand with the saying: "a fine fowl perches only on a fine tree."

D. 20:00–7:00 Resting

▲ Fig.4–18 A green peacock is resting in a tree before the dawn (Photo by the green peafowl research team of Yunnan University using IR camera)

③ Annual activity rhythm .We analyze 1378 independent and effective photos of green peafowl as taken by the green peafowl research team of Yunnan University between January to December in 2017, and calculate the activity intensity of green peafowl for each period (Table 4–2).

Table 4–2 Relative activity intensity of wild green peafowl in each month of 2017 in Xinping County

Month	Independent and effective photos	Relative activity intensity
1	43	3.12%
2	343	24.89%
3	333	24.17%
4	262	19.01%
5	73	5.30%
6	105	7.62%
7	115	8.35%
8	72	5.22%
9	1	0.07%
10	11	0.80%
11	6	0.44%
12	14	1.02%
Total	1378	100%

The annual activity rhythm Fig.4–19 illustrates that green peafowl move around frequently between February and April, while being less active in May. However, they become active again in June and July until another drop in activity is seen from August to January of the next year. Here, a slight twin-peaks model emerges in the annual activity rhythm of green peafowl. Based on images captured by IR cameras, researchers have found that green peafowl prepare for reproduction and start their courting rituals between February and April. Then, green peahens spawn and incubate in May and brood from June to August when they take fledglings out for foraging (Fig.4–19).

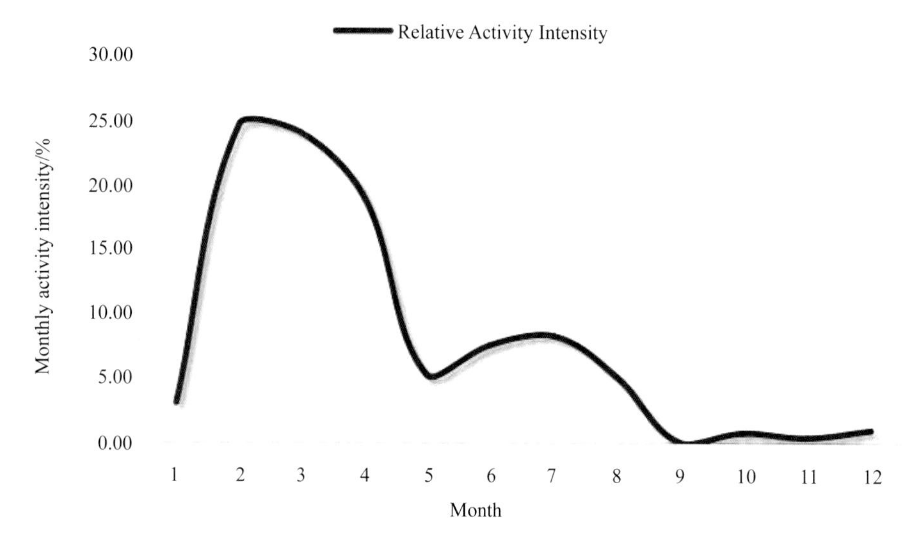

▲ Fig.4–19 Annual activity rhythm of wild green peafowl in Xinping County in 2017

4.3 Relationship between Green Peafowl and Other Species

4.3.1 Symbiosis

(1) Daily activity rhythms of green peafowl (*Gallus gallus*), red junglefowl (*Lophura nycthemera*), silver pheasant and lady amherst's pheasant (*Chrysolophus amherstiae*).

Green peafowl, along with *Red junglefowl*, *Silver pheasant*, and *Lady amherst's pheasant* all belong to the taxonomic family Phasianidae. Through the observation of independent photos taken by IR cameras documenting the activity periods of these four pheasants(Table 4–3), it can be observed that all four animals are diurnal pheasants, only moving around in the day. Green peafowl, *Red junglefowl* and *Silver pheasant* have two activity peaks, while *Lady amherst's pheasant* have several daily activity peaks (Fig.4–20).

Table 4–3 Daily activity intensity data of green peafowl, red junglefowl, silver pheasant and lady amherst's pheasant

Period	Green peafowl		*Red junglefowl*		*Silver pheasant*		*Lady amherst's pheasant*	
	Independent and effective photos	RAI	Independent and effective photos	RAI	Independent and effective photos	RAI	Independent and effective photos	RAI
6	63	4.57%	11	2.06%	0	0	0	0
7	163	11.83%	101	18.88%	5	12.20%	5	12.82%
8	149	10.81%	75	14.02%	9	21.95%	4	10.26%
9	119	8.64%	29	5.42%	2	4.88%	6	15.38%
10	77	5.59%	25	4.67%	0	0	3	7.69%
11	67	4.86%	18	3.36%	0	0	4	10.26%
12	55	3.99%	18	3.36%	2	4.88%	3	7.69%
13	43	3.12%	17	3.18%	0	0	5	12.82%
14	50	3.63%	24	4.49%	0	0	0	0
15	50	3.63%	24	4.49%	4	9.76%	1	2.56%
16	75	5.44%	28	5.23%	3	7.32%	3	7.69%
17	168	12.19%	31	5.79%	10	24.39%	3	7.69%
18	209	15.17%	99	18.50%	5	12.20%	2	5.13%
19	84	6.10%	35	6.54%	0	0	0	0
20	6	0.44%	0	0	1	2.44%	0	0

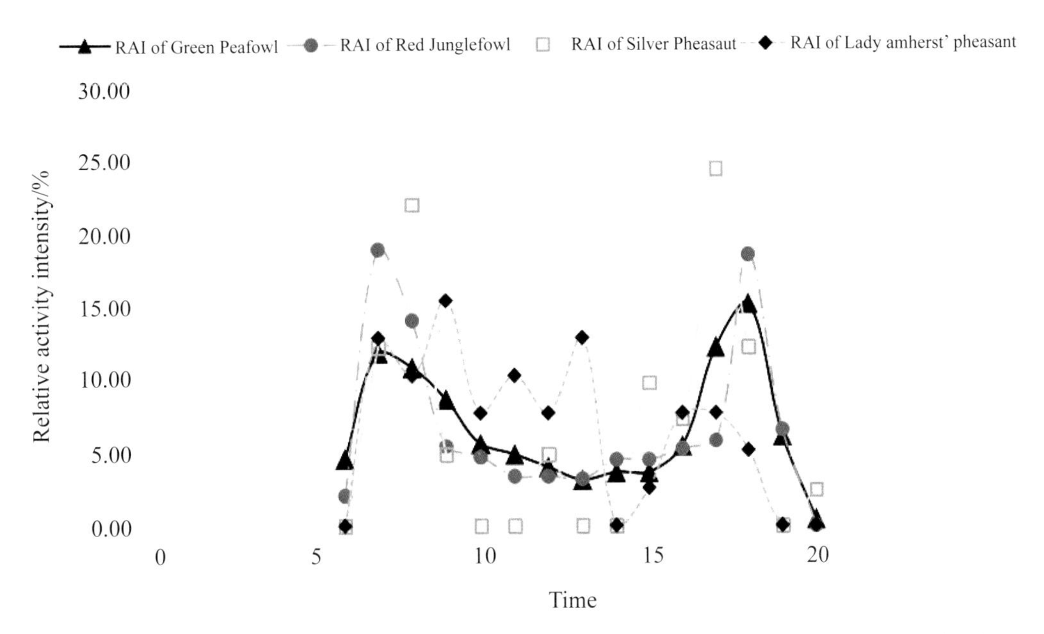

▲ Fig.4–20 Daily Activity rhythms of Green Peafowl, Red junglefowl, Silver pheasant and Lady amherst's pheasant in Xinping County

Red junglefowl move around from 6:00–20:00, the daily activity of which shows significant peaks and bottoms. Specifically, the activity peaks occur in the period from 7:00–9.00 and at 18:00, presenting a dual-peaks model in which maximum values of relative activity intensity happen between 7:00 and 18:00 respectively.

Silver pheasant move around from 7:00–21:00, and also show significant peaks and valleys in their activity. In particular, activity peaks occur in the period from 7:00–9:00 and 17:00–18:00, also presenting a dual-peaks model in which maximum values of relative activity intensity happen at 8:00 and 17:00 respectively.

Lady amherst's pheasant are active between 7:00–19:00, with a concentration of high activity between 7:00–14:00, before a dip is seen from 15:00–19:00. The activity peaks occur at 9:00, 13:00, 16:00, and 17:00.

Green peafowl share the same activity peaks (7:00 and 18:00) with *Red junglefowl*, while those of *Silver pheasant* and *Lady amherst's pheasant* are at 8:00 and 17:00 as well as at 9:00 respectively. From a time perspective, the primary period of activity of the green peafowl and *Red junglefowl* is the same, while compared with *Silver pheasant* their activity is one hour later in the morning and one hour earlier in the afternoon. The activity peak of *Lady amherst's pheasant* in the morning is two hours later than that of green peafowl and *Red junglefowl* but one hour earlier than that of the *Silver pheasant*.

① A peafowl is foraging with a *Red junglefowl* (Fig.4–21). Analysis and statistics of effective photos taken in different periods show the activities of green peafowl, *Red junglefowl, Silver pheasant* and Lady amherst's pheasant, they all demonstrate significant rhythms. The first three pheasants have an activity peak in the morning and afternoon, while *Lady amherst's pheasant* perform most of their movements in the morning, afternoon, and at noon. They are sympatric species and share similar patterns in regards to how they utilize resources and habitats and competing resources internally, however, their activities peak differently. The activities of green peafowl and *Red junglefowl* peak at 7:00, and *Silver pheasant*, at 8:00, which is an hour later. For *Lady amherst's pheasant*, they are most active at 9:00, two hours later than green peafowl and *Red junglefowl* and one hour later than *Silver pheasant*. In the evening, the activity peak of green peafowl and *Red junglefowl* is 18:00, and that of *Silver pheasant* is 17:00, an hour earlier. Nevertheless, the activity peak of *Lady amherst's pheasant* at noon is 13:00, and at 16:00 and 17:00 in the evening, all deviating from those of the green peafowl, *Red junglefowl*, and *Silver pheasant*. This deviation in their activity peaks leads to the separation of the ecological niches of sympatric pheasants who share similar eating patterns and foraging methods and thus reduces the intensity of interspecific competition.

▲ Fig.4–21　A green peafowl is foraging with a red junglefowl (Photo by the green peafowl research team of Yunnan University using IR camera)

② Green peafowl's symbiotic species——*Red junglefowl* (Fig.4–22)

▲ Fig.4–22 Red junglefowl is foraging (♂) (Photo by the green peafowl research team of Yunnan University using IR camera)

③ Green peafowl's symbiotic species——*Silver pheasant* (Fig.4–23–Fig.4–24)

▲ Fig.4–23 Silver pheasant (1♂1♀) (Photo by the green peafowl research team of Yunnan University using IR camera)

▲ Fig.4–24 Silver pheasant (♂) (Photo by the green peafowl research team of Yunnan University using IR camera)

Red junglefowl, *Silver pheasant, Chrysolophus* and *amherstiae* share similar food types and living environments with green peafowl, so there is food competition between wild green peafowl and these birds.

④ Green peafowl's symbiotic species——*Lady amherst's pheasant* (Fig.4–25–Fig.4–26)

▲ Fig.4–25 Sub–adult Lady amherst's pheasant (♂) (Photo by the green peafowl research team of Yunnan University using IR camera)

▲ Fig.4–26 Sub–adult Lady amherst's pheasant (♂) (Photo by the green peafowl research team of Yunnan University using IR camera)

Lady amherst's pheasants are wildlife under second-class national protection that share similar living environments with green peafowl and thus experience food and habitat competition with *Lady amherst's pheasant* and green peafowl.

Paired species matching coefficients represent the probability of green peafowl to appear together with their symbiotic species, and the matching coefficient between green peafowl and *Sus scrofa* is high, meaning that these two species are highly connected. In fact, the relationship between green peafowl and *Sus scrofa* is not obligatory but mutually beneficial, that is, the two species obtain benefits from each other when living together, but can still survive even when separated. Images from IR cameras and on-site observations both demonstrate this result. On one hand, when *Sus scrofa* dig tree roots, soil faunas are dug out as well, offering great convenience to green peafowl when foraging soil faunas. On the other hand, green peafowl are vigilant, and stand as sentry for *Sus scrofa* when humans approach or if danger is present. The two species thus are highly correlated and it's more probable for them to emerge together.(Fig.4–27–Fig.4–28)

▲ Fig.4–27 Green peafowl's symbiotic species——Sus scrofa (Photo by the green peafowl research team of Yunnan University using IR camera)

▲ Fig.4–28 A green peafowl is foraging in places where Sus scrofa have dug (Photo by the green peafowl research team of Yunnan University using IR camera)

The point correlation coefficient (PCC) is divided into 8 periods including $0 \leqslant PCC < 0.3$, $0.3 \leqslant PCC < 0.5$, $0.5 \leqslant PCC < 0.7$, $PCC \geqslant 0.7$, $-0.3 \leqslant PCC < 0$, $-0.5 \leqslant PCC < -0.3$, $-0.7 \leqslant PCC < -0.5$ and $-1 \leqslant PCC < -0.7$ to demonstrate interspecific associations degrees. According to the *PCC* values, green peafowl, *Gallus gallus,Psilopogon asiaticus, Strix aluco, Streptopelia orientalis, Urocissa erythrorhyncha, Dicrurus macrocercus, Callosciurus erythraeus, Capricornis sumatraensis* and *Muntiacus muntjak* are correlated negatively and insignificantly ($0.3 \leq PCC < 0$); while green peafowl, *Silver pheasant, Lady amherst's pheasant, Syrmaticus humiae, Chalcophaps indica, Myophonus caeruleus, Garrulax pectoralis, Macaca mulatta, Sus scrofa, Prionailurus bengalensis* and *Lepus comus* are correlated positively but insignificantly ($0 \leq PCC < 0.3$).

4.3.2 Predatory Relationships

Green peafowl possiblebly share predator prey relationships with *Prionailurus bengalensis*, black bears and other carnivores. Wild adult green peacocks have large bodies, are very vigilant, and can move quickly, all of which greatly reduces their probability of being caught by predators. Juvenile green peafowl, however, are common prey since wild green peafowl nest in such a simple manner that some carnivores are able to steal their eggs. Accordingly, green peafowl abandon their nests or don't give birth to as many fledglings (Fig.4–29).

Crested goshawks mainly prey upon small birds and reptiles, posing a threat to juvenile green peafowl.

▲ Fig.4–29 A Crested Goshawk is found near green peafowl activity sites (Photo by the green peafowl research team of Yunnan University using IR camera)

Note: The crested eagle preys on small birds and reptiles, which can be a threat to the green peafowl.

4.3.3 Symbiotic Relationships

There is no significant food competition between *Muntiacus muntjak* (Fig.4–30, Fig.4–31) and green peafowl since they forage at different times. Research by Wang Fang, etc. (2018) show that the interspecific relationship between these two species is correlated negatively and insignificantly .

▲ Fig.4–30 A Muntiacus muntjak is found near green peafowl activity sites (Photo by the green peafowl research team of Yunnan University using IR camera)

▲ Fig.4–31 Green peafowl's symbiotic species——Muntiacus muntjak (Photo by the green peafowl research team of Yunnan University using IR camera)

4.3.4 Analysis on Inter-species Associations between Wild Green Peafowl in Xinping County and Symbiotic Birds and Beasts

(1) Interspecific Associations Relationship Methodology

Animal's interspecific associations and relationships are linked to species selection and use of living environments, which means there is mutual correlation of animals in spatial distribution. Interspecific association degrees are calculated in a 2 × 2 contingency table (Table 4–4). Each IR camera is regarded as a quadrat and animals that have been recorded are recognized as animals present. Researchers calculate numerical values according to the contingency table. In this research, green peafowl are the target species. Through counting the occurrence of green peafowl and symbiotic species in the quadrat, researchers discuss and analyze the interspecific associations relationships between green peafowl and other symbiotic animals.

Table 4–4 2×2 Contingency Table

		Certain Symbiotic Species		Σ
		Quadrat Present	Quadrat Absent	
Green Peafowl	Quadrat Present	*a*	*b*	*a*+*b*
	Quadrat Absent	*c*	*d*	*c*+*d*
	Σ	*a*+*c*	*b*+*d*	*a*+*b*+*c*+*d*

$$X^2 = \frac{N\left[|ad-bc|-\left(\frac{N}{2}\right)\right]^2}{(a+b)(a+c)(b+c)(c+d)}$$

Verification formula of Interspecific Associations:

In the formula, N represents the total sampling number. When $x^2<3.841$, the interspecific associations are insignificant; when $3.841<x^2<6.635$, associations are significant. When $x^2>6.635$, there are highly significant ecological associations. Meanwhile, the interspecific associations can be determined on the basis of the values of ad-bc. If the values are greater than 0, there are positive associations, but if these values are less than 0, then there are negative associations. Positive associations represent a high encounter chance and an overlap

of the environment niches of two species, while negative associations represent a low encounter chance and point to the occupation of different foraging spaces.

Calculation of interspecific associations:

①Point correlation coefficient (*PCC*) index. The PCC is adopted to represent the interspecific associations degrees of green peafowl and symbiotic birds and beasts. When the *PCC* is less than 0, there are negative associations; when the *PCC* is greater than 0, there are positive associations. The larger the absolute values are, the higher the associations degrees are. When PCC is 0, it means the species are completely independent.

$$PCC = \frac{ad - bc}{\sqrt{(a+b)(a+c)(b+d)(c+d)}}$$

Note: a represents the number of quadrat in which both animals occur; b represents the number of quadrat in which green peafowl occur but symbiotic species do not; c represents the number of quadrat in which symbiotic species occur but green peafowl do not; d represents the number of quadrat in which neither animal occurs.

②Paired species matching coefficients (*PC*).

PC coefficients are adopted to represent the interspecific associations degrees and the chance for green peafowl and symbiotic birds and beasts to emerge together. When there is "no correlation," the value is 0, and when there is "maximum correlation," the value is 1.

Coefficient of community:

$$PC = \frac{a}{a+b+c}$$

(2) Interspecific association degree between green peafowl and symbiotic birds and beasts

According to the data from IR cameras, a 2 × 2 contingency table is made (Table 4–5). 42 IR camera sites have captured images of green peafowl and symbiotic birds and beasts. Based on these, researchers have calculated the interspecific associations degrees between green peafowl and 27 symbiotic birds and beasts.

Table 4–5 The interspecific association relationship between green peafowl and associated birds and animals

Species	*a*	*b*	*c*	*d*	x^2	*PCC*	*PC*
Lady amherst's pheasant	10	18	0	14	4.741	0.395	0.357
ophura nycthemera	8	20	0	14	3.262	0.343	0.286
Red junglefowl	21	7	1	13	14.616	0.640	0.724
Copsychus malabaricus	1	27	0	14	0.128	0.110	0.036
Enicurus maculates	1	27	0	14	0.128	0.110	0.036
Megalaima virens	1	27	0	14	0.128	0.110	0.036
Syrmaticus humiae	0	28	1	13	0.128	−0.221	0
Dicrurus macrocercus	0	28	1	13	0.128	−0.221	0
Garrulax pectoralis	4	24	1	13	0.028	0.104	0.138
Turdus dissimilis	5	23	0	14	1.391	0.260	0.179
Urocissa erythrorhyncha	1	27	3	11	1.692	−0.287	0.032
Strix aluco	0	28	1	13	0.128	−0.221	0
Megalaima asiatica	0	28	1	13	0.128	−0.221	0
Chalcophaps indica	3	25	0	14	0.404	0.196	0.107
Copsychus saularis	2	26	0	14	0.066	0.158	0.071
Streptopelia orientalis	0	28	4	10	5.837	−0.459	0
Pica pica	1	27	0	14	0.128	0.110	0.036
Myophonus caeruleus	4	24	0	14	0.863	0.229	0.143
Prionailurus bengalensis	14	14	5	9	0.300	0.135	0.424
Muntiacus vaginalis	19	9	5	9	2.734	0.306	0.576
Callosciurus erythraeus	7	21	0	14	2.593	0.316	0.250
Paguma larvata	1	27	1	13	0.066	−0.079	0.034
Macaca mulatta	4	24	0	14	0.863	0.229	0.143
Tamiops macclellandii	2	26	0	14	0.066	0.158	0.071
Sus scrofa	10	18	5	9	0.117	0	0.303
Lepus comus	1	27	1	13	0.066	−0.079	0.034
Atherurus macrourus	9	19	0	14	4.771	0.369	0.321

Note: x^2: Interspecific associations index; *PCC*: Point correlation coefficient; *PC*、*OI*: Paired species matching coefficients

The interspecific association coefficient x^2 is divided into three levels: when $x^2>6.635$, the interspecific associations are extremely significant; when $x^2<3.841$, associations are insignificant; and when $3.841<x^2<6.635$, associations are significant. Verification of the interspecific associations between green peafowl and symbiotic birds and beasts demonstrate:

① There are extremely significant ecological associations ($x^2\geqslant6.635$) between green peafowl and *red junglefowl* (14.616);

② There are significant ecological associations ($3.841<x^2<6.635$) between green peafowl and *Streptopelia orientalis* (5.837), *Atherurus macrourus* (4.771) and *Lady amherst's pheasant* (4.741);

③ There are insignificant ecological associations ($x^2<3.841$) between green peafowl and *Silver pheasant* (3.262), *Muntiacus vaginalis* (2.734), *Callosciurus erythraeus* (2.593), *Urocissa erythrorhyncha* (1.692), *Turdus dissimilis* (1.391), *Myophonus caeruleus* (0.863), *Macaca mulatta* (0.863), *Chalcophaps indica* (0.404), *Prionailurus bengalensis* (0.3), *Copsychus malabaricus* (0.128), *Enicurus maculates* (0.128), *Megalaima virens* (0.128), *Syrmaticus humiae* (0.128), *Dicrurus macrocercus* (0.128), *Strix aluco* (0.128), *Megalaima asiatica* (0.128), *Pica pica* (0.128), *Sus scrofa* (0.117), *Copsychus saularis* (0.066), *Paguma larvata* (0.066), *Tamiops macclellandii* (0.066), *Lepus comus* (0.066) and *Garrulax pectoralis* (0.028).

4.3.5 Point Correlation Coefficient

The point correlation coefficients are divided into 8 periods including $0\leqslant PCC<0.3$, $0.3\leqslant PCC<0.5$, $0.5\leqslant PCC<0.7$, $PCC\geqslant0.7$, $-0.3\leqslant PCC<0$, $-0.5\leqslant PCC<-0.3$, $-0.7\leqslant PCC<-0.5$ and $-1\leqslant PCC<-0.7$ to demonstrate the interspecific associations degrees. The *PCC* numerical values demonstrate that:

① There are positive correlations between green peafowl and *Atherurus macrourus* (0.64), *Red junglefowl* (0.395), *Garrulax pectoralis* (0.369), *Streptopelia orientalis* (0.343), *Pica pica* (0.316), *Megalaima asiatica* (0.306), *Myophonus caeruleus* (0.26), *Dicrurus macrocercus* (0.229), *Copsychus saularis* (0.229), *Copsychus malabaricus* (0.196), *Enicurus maculates* (0.158), *Paguma larvata* (0.158), *Strix aluco* (0.135), *Lady amherst's pheasant*

(0.11), *Silver pheasant* (0.11), *Muntiacus vaginalis* (0.11), *Syrmaticus humiae* (0.11) and *Turdus dissimilis* (0.104), in which the correlation between green peafowl and *Atherurus macrourus* is the strongest.

② There is no correlation between green peafowl and *Tamiops macclellandii* (0).

③ There are negative correlations between green peafowl and *Sus scrofa* (−0.079), *Lepus comus* (−0.079), *Callosciurus erythraeus* (−0.221), *Urocissa erythrorhyncha* (−0.221), *Chalcophaps indica* (−0.221), *Prionailurus bengalensis* (−0.221), *Macaca mulatta* (−0.287) and *Megalaima virens* (−0.459), in which the correlation between green peafowl and *Megalaima virens* is the strongest.

4.3.6 Paired Species Matching Coefficients

The numerical values of paired species matching coefficients are divided into 4 sections to demonstrate interspecific associations degrees, including $PC<0.3$, $0.3\leqslant PC<0.5$, $0.5\leqslant PC<0.7$ and $PC\geqslant 0.7$. The *PC* numerical values show that:

Green peafowl have the strongest correlation with *Garrulax pectoralis* (0.724) and the two species are most probable to encounter each other; next comes *Sus scrofa* (0.576), *Tamiops macclellandii* (0.424), *Atherurus macrourus* (0.357), *Megalaima virens* (0.321), *Prionailurus bengalensis* (0.303), *Red junglefowl* (0.286), *Lepus comus* (0.250), *Copsychus malabaricus* (0.179), *Turdus dissimilis* (0.143), *Urocissa erythrorhyncha* (0.143), *Copsychus saularis* (0.138), *Lady amherst's pheasant* (0.107), *Silver pheasant* (0.071), *Chalcophaps indica* (0.071), *Streptopelia orientalis* (0.036), *Pica pica* (0.036), *Megalaima asiatica* (0.036), *Syrmaticus humiae* (0.036), *Callosciurus erythraeus* (0.034), *Macaca mulatta* (0.034) and *Enicurus maculates* (0.032). However, green peafowl do not possess a correlation with *Myophonus caeruleus* (0), *Dicrurus macrocercus* (0), *Paguma larvata* (0), *Strix aluco* (0) and *Muntiacus vaginalis* (0).

4.3.7 Discussions on the inter-specific relationships between green peafowl and symbiotic birds and beasts

The interspecific associations index x^2 represents the encounter chance and any overlaps of environment niches between two species. Paired species with significant or extremely significant positive associations tend to live in similar environments or share similar demands in regard to their habitat environments or ecological niches. The verification of the interspecific association relationships between green peafowl and these 27 symbiotic birds and beasts shows that green peafowl are significantly correlated with *red junglefowl* ecologically, which means both species share similar living spaces and patterns of selecting habitats. Their encounter and ecological overlap probabilities are high. These results correspond with the activity rhythms of the two species. In addition, green peafowl are significantly correlated with *Streptopelia orientalis, Lady amherst's pheasant* and *Atherurus macrourus* in their ecology and are very likely to encounter them, while green peafowl are not correlated with other birds and beasts, and are not likely to encounter those species.

The point correlation index (*PCC*) shows that there are positive correlations between green peafowl and *Atherurus macrourus*, *Red junglefowl, Garrulax pectoralis, Streptopelia orientalis*, *Pica pica, Megalaima asiatica, Myophonus caeruleus, Dicrurus macrocercus, Copsychus saularis, Copsychus malabaricus, Enicurus maculates, Paguma larvata, Strix aluco,Lady amherst's pheasant, Silver pheasant, Muntiacus vaginalis, Syrmaticus humiae and Turdus dissimilis*, in which the correlation between green peafowl and *Atherurus macrourus* is the strongest and the values decrease successively. There is no correlation between green peafowl and *Tamiops macclellandii*. Finally, there are negative correlations between green peafowl and *Sus scrofa, Lepus comus, Callosciurus erythraeus, Urocissa erythrorhyncha, Chalcophaps indica, Prionailurus bengalensis, Macaca mulatta* and *Megalaima virens*, in which the correlation between green peafowl and *Megalaima virens* is the strongest and the values increase successively.

Paired species matching coefficients (*PC*) illustrate the probability for two species to emerge together, where the codomain is [0, 1]. The closer the value is to 1, the closer the correlation is between two species. *PC* values show that green peafowl are most closely correlated with *Garrulax pectoralis*, followed by *Sus scrofa, Tamiops macclellandii, Atherurus macrourus, Megalaima virens, Prionailurus bengalensis, Red junglefowl, Lepus comus, Copsychus malabaricus, Turdus dissimilis, Urocissa erythrorhyncha, Copsychus saularis, Lady amherst's pheasant, Silver pheasant, Chalcophaps indica, Streptopelia orientalis, Pica pica, Megalaima asiatica, Syrmaticus humiae, Callosciurus erythraeus, Macaca mulatta* and *Enicurus maculates*. Green peafowl, however, do not have a correlation with *Myophonus caeruleus, Dicrurus macrocercus, Paguma larvata, Strix aluco* and *Muntiacus vaginalis*.

Positive correlations occur for two reasons. The first is the dependence of one species on another and the second is the similar adaptation and response to heterogeneous environmental conditions by more than one species. Conversely, negative correlation occurs because of the mutual repulsion of species in resource competition. Among species positively correlated with green peafowl, there are no species that depend on or are dependent on the latter, thus they have parallel adaptations and share responses to environmental conditions in heterogeneous environments; but species negatively correlated with green peafowl repel the latter due to reasons related to resource competition.

5

Behavioral Characteristics of Green Peafowl

5.1 Foraging Behavior

Typical foraging behavior of green peafowl entails directly pecking at fallen fruits and seeds found on the ground with its beak. Green peafowl will also peck at tender leaves from low-growing plants, dig out grass roots, and search for insects under fallen leaves with its beak and claws (Fig.5–1–Fig.5–3). Feeding occupies the largest proportion (51.82%) of a typical green peafowl's time, and takes place primarily between the hours of 7 to 12 a.m. and 6 to 8 p.m. (Yang Xiaojun et al., 2000).

▲ Fig.5–1 A green peafowl foraging for food (Photo by the green peafowl research team of Yunnan University using IR camera)

▲ Fig.5–2　A green peafowl is fetching food (Photo by the green peafowl research team of Yunnan University using IR camera)

▲ Fig.5–3　A green peafowl is looking for food (Photo by the green peafowl research team of Yunnan University using IR camera)

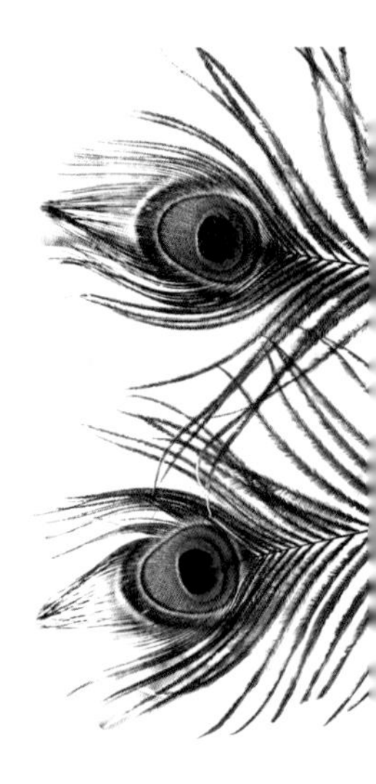

5.2 Community Behavior

Green peafowl are generally polygamous, and usually form groups of about 5 to 10 peafowl. There is only one adult peacock in each group while the rest are family members composed of peahens or subadults (Fig.5–4–Fig.5–7). The size of each group will also vary with the seasons. During the winter these groups tend to be larger while during other seasons their family activities are scattered .

▲ Fig.5–4 The green peacocks and peahens are looking for food together. (Photo by the green peafowl research team of Yunnan University using IR camera)

▲ Fig.5–5 A green peafowl is eating (Photo by the green peafowl research team of Yunnan University using IR camera)

▲ Fig.5–6 Green peahens often forage together (Photo by the green peafowl research team of Yunnan University using IR camera)

Adult male green peafowl are fierce and aggressive. Fights between males during the breeding season happen mainly through the display of courtship by unfolding the tail screen, which can sometimes be very intense.

▲ Fig.5–7 The green peacocks and peahens are foraging together (Photo by the green peafowl research team of Yunnan University using IR camera)

The breeding season of green peafowl living in the banks of Shi Yang River at the foot of Ailaoshan and begins around mid to late February every year. At that time, male courting calls can be heard in the valley and on both sides of the mountains, one after another, as they resound and echo all over the landscape —— this is the green peacock's claim to sovereignty. If other green peafowls chirp in its territory, it will approach the intruder, chirping to warn him and drive him away. If the intruder refuses to leave, there will be a fierce fight for the same peahen.

In a small green peafowl family, there is only one adult peacock who is responsible for the protection of other peafowl and the reproduction of the entire group. Males of different families all have their own living areas (Fig.5–8).

▲ Fig.5–8 A subadult green peacock is foraging alone (Photo by the green peafowl research team of Yunnan University using IR camera)

In addition to foraging together, green peafowl will help each other preen their feathers while at rest (Fig.5–9).

▲ Fig.5–9 A group of green peahens are preening together (Photo by the green peafowl research team of Yunnan University using IR camera)

5.3 Reproductive Behavior

The breeding season of green peafowl spans from March to June. The rutting green peacock performs graceful courtship displays. It excitedly spreads its tail coverts and supports them on its raised, fanned-out tail feathers, swinging and turning to the left and right. While quickly shaking its tail coverts, beautiful eyespots (ocelli) shine gracefully. This kind of mesmerizing courtship performance is commonly known as a "peafowl display" (Fig.5–10–Fig.5–11). The peacock sends out its courting signal through the performance of this graceful dance. Then, with its head stretching out, it approaches and circles around the peahens, excitedly unfolding and displaying the full magnificence of its tail feathers. The peacock usually displays its feathers anywhere up to 15 times to the peahen, with different lengths of time being used for each showing, varying from a few seconds up to two hours. When performing its display, the peacock usually "boasts" to the peahen through its body rotation, its swaying to the left and right, and the colorful reflections of its ocelli peahens. The peacock shakesits tail at a high frequency and often stamps its feet to arouse the attention of the peahen that stands or passes in front of it. The mating of peafowl is a behavioral chain in which the peacock courts, and the peahen responds by lying down and crawling, then the peacock pecks at and steps onto the peahen's head in the process of mating. After its first successful mating, the peacock continues to copulate instead of immediately closing its tail.

▲ Fig.5–10 An adult green peacock is displaying its tail feathers (Photo by the green peafowl research team of Yunnan University using IR camera)

▲ Fig.5–11 A group of green peafowl in breeding season (Photo by the green peafowl research team of Yunnan University using IR camera)

The peahen hatches eggs alone. The incubation period of the green peafowl ranges from 27 to 30 days, and it takes a hatching chick 30 hours on average to peck apart its shell and finally come out. The chick's squeaks and the sounds of its pecking last between 1 to 2

days before fully hatching out. The sound of a chick hatching is weak at first, but gradually grows stronger and increases in frequency. The chick pecks out a small hole at the transverse diameter of 1 ~ 1.5 cm from the blunt end of the egg, and then pecks out a crack near the middle of the shell that divides the egg into a larger and a smaller end before it finally removes the shell entirely and emerges from the egg (Fig.5–12）.

▲ Fig.5–12 A green peahen is hatching eggs (Photo by the green peafowl research team of Yunnan University using IR camera)

After mating season, the nesting and incubation period of the chick begins from March to April. Green peafowl usually nest in the hollow ground covered with branches and weeds, and is surrounded by shrubs which provide vital concealment. The nest is extremely simple and crude. The green peafowl lays 3 to 6 eggs at a time. These oval eggs are milky white or yellow, smooth, and spot-free (Fig.5–13）.

▲ Fig.5–13 The nest and eggs of green peafowl (Photo by the courtesy of Wang Fang)

The chicks are precocial and can follow the peahen to forage for food after hatching. It takes the young peacock three years to grow gorgeous tail feathers (Fig.5–14）.

▲ Fig.5–14 A green peahen is rearing chicks (Photo by the green peafowl research team of Yunnan University using IR camera)

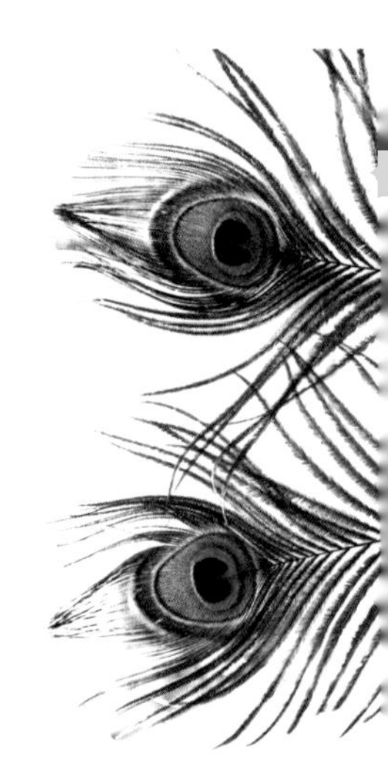

5.4 Evading Predators

Peafowl are very alert by nature, and are fast on the ground due to their strong feet. Although they are not particularly adept at flying, they can fall at a faster speed. Peafowl are social animals and seldom act alone. They gather in larger groups in autumn and winter. When a wild green peafowl encounters its natural predator, it will scream loudly before quickly fleeing or glide a short distance to avoid its enemy. Most of the time, it can escape quickly with its long, sturdy legs.

▲ Fig.5-15 A green peacock is looking around (Photo by the green peafowl research team of Yunnan University using IR camera)

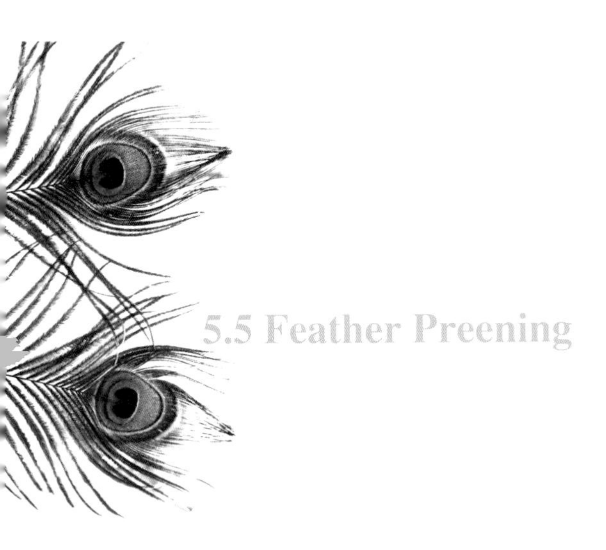

5.5 Feather Preening

Grooming is an important activity in the daily life of green peafowl. When they rest, they will choose a safe place to groom their feathers with their beaks and apply the oil from their tail fat glands to keep their feathers shiny. Both male and female adults pay special attention to grooming, especially the male adult green peafowl during the breeding period (Fig.5-16-Fig.5-17).

▲ Fig.5-16 A green peafowl is preening (Photo by the green peafowl research team of Yunnan University using IR camera)

▲ Fig.5–17 A green peafowl is preening (Photo by the green peafowl research team of Yunnan University using IR camera)

5.6 Unfolding tail screens, walking, observing and alerting behavior

The behavior of green peafowl in the wild is very diverse. Because they are very alert, only the photos and videos captured by the infrared camera can be used to obtain a fragment of their behavior in the field. The unfolding of the tail screen, walking, observation and vigilance of Green Peafowl are shown in Fig.5-18–Fig.5-21.

▲ Fig.5-18 The peafowl is unfolding tail screens (Photo by the green peafowl research team of Yunnan University using IR camera)

▲ Fig.5-19　The peafowl is walking (Photo by the green peafowl research team of Yunnan University using IR camera)

▲ Fig.5-20　The green peacock in the post-breeding period take off the tail coverts (Photo by the green peafowl research team of Yunnan University using IR camera)

▲ Fig.5–21 A solitary green peafowl at alert (Photo by the green peafowl research team of Yunnan University using IR camera)

6 Current Research and Conservation Status of Wild Green Peafowl in China

The Yunnan Provincial Committee of the Communist Party of China, the People's Government of Yunnan Province, the Yunnan Forestry and Grassland Administration, and the Department of Ecology and Environment of Yunnan Province have done a lot of work and made important headway for the conservation and management of green peafowl. Firstly, the planning and establishment of conservation areas is underway. Among the various levels of nature reserves established in Yunnan, there are 15 reserves in which green peafowl are distributed in an effort to conserve and manage the species as one of the primary focuses of conservation. Secondly, green peafowl are included among the list of species with extremely small populafion. In 2007, the Yunnan Forestry Department took the lead in proposing the conservation of species with extremely small populations in the country and named green peafowl as one of the 20 key species for priority conservation. Thirdly, population surveys and monitoring have been carried out. Green peafowl were investigated in the first and second national wildlife surveys and in 2017 and 2018, the Forestry Department of Yunnan Province and the Kunming Animal Research Institute of the Chinese Academy of Sciences launched the "Investigation and Evaluation of Green Peafowl Populations and its Habitat in the Upper Reaches of the Yuanjiang River" and the "Investigation of Green Peafowl Resources in the Province" projects, thereby further investigating the status of green peafowl numbers in Yunnan Province as well as laying a foundation for the scientific conservation of green peafowl. Fourthly, small, co-managed nature reserves of green peafowl habitats have been established. Led by the Yunnan Forestry and Grassland Administration and funded by the Alashan League, this project was implemented in 2018 in Yao Village, Xinping County. This undertaking has been guided by experts and supported by the relevant departments at every step along the way, and has seen marked results, providing a basis for exploring a model of green peafowl conservation and management (Fig 6–1—Fig.6–2).

▲ Fig.6–1　Conducting the investigation of green peafowl distribution (Photo by the courtesy of Wang Fang)

▲ Fig.6–2　Conducting green peafowl habitat investigations (Photo by the courtesy of Chen Mingyong)

6.1 Previous Research of Wild Green Peafowl in China

6.1.1 Changes in the Distribution of Wild Green Peafowl in Chinese History

There were few studies on Chinese wild green peafowl in history, and the first documented record was verified by Wen Huanran et al. in 1980. They discovered that green peafowl were once found throughout Hunan, Hubei, Sichuan, Guangdong, Guangxi, Yunnan and other provinces and regions. By the beginning of the 21st century, the green peafowl had disappeared in other provinces outside Yunnan and northeastern Yunnan. The distribution areas of green peafowl in China had shrunk to the western, central, and southern parts of Yunnan province.

6.1.2 Population

According to surveys conducted via letters and field surveys by Wen Xianji et al. (1995) from 1991 to 1993, the population of green peafowl was at its peak before the 1960s, after which it saw a sharp drop due to continuous habitat loss and excessive hunting. It was estimated that the population of green peafowl in Yunnan was about 800 to 1100, as part of an earlier and more comprehensive report on the population of green peafowl in Yunnan. A survey of green peafowl in Xishuangbanna, Yunnan, was conducted from 1994 to 1995, and the results showed that the population of green peafowl in Xishuangbanna was somewhere between 19 and 25. In 1999, the population survey in Chuxiong Yi Autonomous Prefecture was 280(Xu Hui, 1995). In the following decade, there were no research reports on the species' population.

Wen Yunyan et al. (2016) used an IR camera to monitor and plot green peafowl in Konglonghe Prefectural Nature Reserve of Shuangbai, Chuxiong as part of his investigation. The results showed that there were 56 green peafowl in the Konglonghe Natural Reserve.

From April 2014 to June 2017, Kong Dejun and Yang Xiaojun surveyed China's green peafowl resources via interviews, questionnaires and line-transect method. They also made use of plotting, bird calls, and IR cameras to supplement the survey. It showed that the population of green peafowl in China was less than 500 (Kong Dejun and Yang Xiaojun, 2017).

Hua Rong et al. (2018) estimated that the population of green peafowl in China was between 235 and 280 through letters and field surveys from 2015 to 2017 (Table 6–1). In the past two decades, the population of green peafowl in China has decreased drastically, and green peafowl have disappeared in many areas. Even in Xishuangbanna, Yunnan Province, which is dubbed "The hometown of Peafowl," it is difficult to even see traces of the green peafowl.

Table 6–1 Populations of wild green peafowl in China reported over the years

Years	Investigated areas	Population of green peafowl	Data source
1995	Yunnan Province	800–1100	Wei Xianji et al., 1995
1998	Xishuangbanna Prefecture, Yunnan Province	19–25	Luo Aidong and Dong Yonghua, 1998
1999	Chuxiong Prefecture, Yunnan Province	About 280	Xu Hui, 1995
2016	Konglonghe Prefectural Nature Reserve of Chuxiong Prefecture, Yunnan Province	56	Wen Yunyan et al., 2016
2018	Yunnan Province	Less than 500	Kong Dejun and Yang Xiaojun, 2017
2018	Yunnan Province	235–280	Hua Rong et al., 2018

6.1.3 Distribution Changes in History

In Guangdong, Guangxi, Hubei, Hunan, Sichuan, and Yunnan, green peafowl were widely distributed in accordance with historical records. But, at the beginning of this century, green peafowl in northeastern Yunnan and other provinces outside Yunnan all went extinct, and their distribution area has been reduced to southern, central and western Yunnan.

The results of research projects conducted by Wang Zijiang in 1983 demonstrated that green peafowl were spread across Lushui, Tengchong, Yingjiang, Mengding, Xishuangbanna, and Xinping in Yunnan Province. Later, it was reported in 1990 that they were also found in four counties in the Chuxiong Yi Autonomous Prefecture, consisting of Lufeng, Shuangbai, Nanhua, and Yao'an, as well as in Chuxiong City (Wang Zijiang, 1990).

The results of the research on the green peafowl in Yunnan Province conducted by Wen Xianji et al. (1995) from 1991 to 1993 revealed that green peafowl could be found distributed across 34 counties (cities), with two counties of Weixi and Deqin waiting to be confirmed. Five counties which green peafowl had once called home had become devoid of the species completely (Table 6–2).

Yang Xiaojun et al. (1997) conducted surveys of green peafowl in southeastern and northwestern Yunnan Province from 1995 to 1996 and showed that there were only 3 counties in southeastern Yunnan where green peafowl were distributed, including Jianshui, Shiping, and Mile. There were 6 counties (cities) including Wenshan, Mengzi, Jinping, Lüchun, Hekou, and Kaiyuan, where green peafowl had become newly extinct.

Luo Aidong et al. (1998) surveyed green peafowl in Xishuangbanna Prefecture from 1994 to 1995 and reported that green peafowl were only distributed in parts of Jinghong City, Menghai County, and Mengla County.

Han et al. investigated green peafowl in Yunnan Province in 2007 and found that in the history of Yunnan Province, green peafowl had once been distributed over 42 areas. In the five counties of Yingjiang, Tengchong, Liuku, Mengzi, and Hekou, green peafowl became extinct in the 1980s, while in the four counties of Mengla, Jinping, Lüchun, and Jianshui, they had vanished completely in the 1990s. At present, green peafowl are found only in 31 counties (cities).

Kong et al. launched a comprehensive survey of green peafowl in Yunnan Province from 2014 to 2017. In the past 30 years, the distribution area of green peafowl in Yunnan Province has shrunk sharply from 11 states (cities), 34 counties, and 127 towns, to 8 states (cities), 22 counties, and 33 towns. There were no green peafowl in Chayu and Motuo counties in the Tibet Autonomous Region. In addition, Yuanjiang and Eshan counties in central Yunnan are newly discovered distribution areas (Table 6–2).

Table 6–2 Distribution information of Chinese green peafowl recorded in previous surveys

Time	The Distribution of Green Peafowl	Data Resource
Before the 20th Century	Hunan, Hubei, Sichuan, Guangdong, Guangxi, Yunnan	Wen Huanran, He Yeheng, 1980
1983, 1990	Green peafowl were distributed in Lushui, Tengchong, Yingjiang, Mengding, Xinping, Lufeng, Shuangbai, Nanhua, Yao'an, Chuxiong City, and Xishuangbanna Prefecture in Yunnan Province.	Wang Zijiang, 1983, 1990
1995	Green peafowl were distributed in 34 counties (cities) in Yunnan Province including Ruili, Longchuan, Luxi (now Mang City), Changning, Longling, Yongde, Zhenkang, Gengma, Cangyuan, Shuangjiang, Yunxian, Lincang (now Linxiang District), Fengqing, Xinping, Pu'er(now Ning'er County), Mojiang, Jingdong, Jinggu, Zhenyuan, Simao (now Simao District), Chuxiong (now Chuxiong City), Shuangbai, Nanhua, Yongren, Yao'an, Lufeng, Jinghong (now Jinghong City), Menghai, Mengla, Weishan, Lüchun, Jinping, Shiping,and Mile. They become extinct in 5 counties (cities) including Yingjiang, Lushui, Tengchong (now Tengchong City), Mengzi, and Hekou.	Wen Xianji et al, 1995
1997	Green peafowl were distributed in 3 counties in Yunnan Province including Jianshui, Shiping, and Mile.They die out in 5 counties including Mengzi, Jinping, Lüchun, Hekou, Kaiyuan, Wenshan, etc.	Yang Xiaojun et al. 1997
2007	Green peafowl were found in 31 counties in Yunnan Province including Weishan, Yongren, Jinghong, Ruili, Longchuan, Luxi, Longling, Changning, Fengqing, Yunxian, Yongde, Zhenkang, Gengma, Cangyuan, Shuangjiang, Lincang, Jingdong, Jinggu, Zhenyuan, Pu'er (now Ning'er), Simao, Menghai, Mojiang, Shiping, Mile, Xinping, Shuangbai, Chuxiong, Lufeng, Nanhua, Yao'an, and other areas.	Han et al., 2007
2018	Green peafowl were distributed in 22 counties in Yunnan Province including Ruili, Longchuan, Longling, Changning, Yongde, Zhenkang, Gengma, Lancang, Jinggu, Jingdong, Ning' er, Mojiang, Jinghong, Yuanjiang, Xinping, Eshan, Chuxiong, Nanhua, Lufeng, Shuangbai, Shiping, and Jianshui.	Kong et al., 2018

From January to June 2017, Wang Fang et al. used IR cameras to continuously monitor wildlife in Xinping County, Yunnan Province. According to statistics, 33 IR cameras have been in operation for a total of 3836 working days, capturing 1853 groups of independently useful photo and video, among which there were 1473 groups of birds and 380 groups of mammals. Thirteen species of birds were captured, belonging to 5 orders, 8 families, and 13 genera, while 8 species of mammals, belonging to 5 orders, 8 families, and 8 genera were also documented. Among them, two animals fell within national level animal protection categories, namely, Pavo muticus and Syrmaticus humiae. There were 5 animals under the second-class state protection—the Gallus gallus, Lophura nycthemera, Chrysolophus amherstiae, Macaca mulatta, and Capricornis milneedwardsii. Green peafowl were listed on the IUCN Red List of Threatened Species. The G-F index was adopted to calculate the diversity index of birds and mammals and quantitatively analyze photos taken by IR cameras to observe the diversity found within species between families and genera. It was found that the G-F index of birds was higher than that of mammals, indicating that the same would hold true in that specific area. Three more diversified species of birds were Pavo muticus, Gallus gallus, and Lophura nycthemera; on the mammal end of things were rodents, Callosciurus erythraeus, and Prionailurus bengalensis (Wang Fang et al., 2018). The interspecific relationships of green peafowl mainly included predator and prey relationships, interspecific competition, parasitism, courtship and so on.

According to the study conducted by Wang Fang et al. (2018) on the diversity of birds and mammals associated with green peafowl and the relationship between them, there were more than 20 species of birds and mammals associated with green peafowl. These birds include the *Gallus gallus*, *Lophura nycthemera Syrmaticus humiae*, *Chrysolophus amherstiae*, *Megalaima asiatica*, *Strix aluco*, *Chalcophaps indica*, *Streptopelia orientalis*, *Myophonus caeruleus*, *Garrulax pectoralis*, *Urocissa erythrorhyncha*, and *Dicrurus macrocercus*, while mammals included the *Muntiacus vaginalis*, *Sus scrofa*, *Macaca mulatta*, *Prionailurus bengalensis*, *Capricornis milneedwardsii*, *Callosciurus erythraeus*, and *Lepus comus*.

While using the Ejia area of the Ailaoshan National Park in Chuxiong (i.e. Ejia Town, Shuangbai County) as her foundation, Jiao Yuanmei et al. (2017) carried out functional zoning based on the potential habitat of the IUCN endangered species of green peafowl. The results showed that: ① The construction of the Ejia area of the Ailaoshan National Park in Chuxiong was of great significance to protect the integrity of the surrounding ecosystems which consisted of mid-montane wet evergreen broad-leaved forest and valley monsoon rain forests in Ailaoshan while also allowing its flagship species—particularly the endangered species of green peafowl—to realize the spatial integration of conservation areas and protected objects; ② Based on the extracted information about altitude, slope, aspect, and vegetation types, the viable habitat and unviable habitats of green peafowl accounted for 20.32% and 79.68% of the total area respectively. Because the potential habitat patches were small and fragmented, those with a spacing of less than 500 meters would be connected, and the core area of the potential habitat could be extracted in line with the density of the connecting line; ③ In protecting the integrity of the ecosystem as well as suitable habitats for endangered species, and under the principle of integrating the original conservation areas with community developments and recreation areas, the Ejia area was divided into core conservation areas, ecological conservation areas, traditional utilization areas, recreational exhibition areas, and human activity areas, accounting for 66.90%, 16.80%, 7.01%, 0.24%, and 9.05% of the allocated land respectively. The conservation and utilization requirements of each functional area were different. In addition, researchers were assigned to green peafowl habitats to study their feeding habits, breeding and reproduction, physiology and genetics, disease and prevention, threats, and conservation management. Research on green peafowl habitats have been carried out in Java Island in Indonesia (Jarwadi et al., 2011) and Vietnam (Briekle, 2002). In China, observations and studies on green peafowl habitats in Jingdong County has also been conducted early (Yang Xiaojun, 2000). Studies on food habits of green peafowl have shown that it was omnivorous, feeding on oak trees, *pyracantha fortuneana*, *Valeriana officinalis* L., legumes, new leaves and flowers of Asteraceae, grass seeds, rice, etc. They also were observed to like eating pears, Magnoliopsida, and also prey on locust, scarab insects, crickets, grasshoppers, small moths, frogs, lizards, etc.(Kuang Bangyu, 1963; Wen Huaran and He Yeheng, 1980; Wang Zijiang, 1983; Kong Dejun and Yang Xiaojun, 2017). Feeding and reproduction were mainly related research conducted regarding its feed composition, hatching methods, temperature, humidity, and moisture during the hatching process (Zhang Chunli, 1995; Wu Jun, 2004; Li Shiqiang et al., 2006; Wang Fenghua et al.,

2007; Zhang Zhong'an, 2008; Zhang Lixia, et al., 2015). Studies on physiology and genetics concerned mainly the composition of their shells (Wang Yulong et al., 2000), the histological observation of their digestive system (He Ping, and Lu Yuyan, 2002; Li Jian et al., 2004), blood physiological indicators (Qi Weiwei et al., 1998; Zhang Yunmei et al., 2003; Zhou Qingping et al., 2011), and on the taxonomic status of green peafowl(Chang Hong et al., 2002; Ke et al., 2004; Zhu Shijie et al., 2004; Bao Wenbin et al., 2006; Ouyang Yina et al., 2009; Duan Yubao et al., 2018). Disease and prevention were mainly concerned with the studies of green peafowl Histomoniasis (Wang Xiumei et al., 2001; Yan Gang and Hou Junli, 2002; Shi Xiaotao et al., 2008; Zhou Wei et al., 2015), coccidiosis (Zhang Lichun and Li Musen, 2008; Li Musen and Cui Zhen'ai, 2009; Chen Jing, 2011), bacterial infection (Zhang Jianhu and Deng Nijuan, 2002; Zhao Hengzhang and Li Junmin, 2004; Hu Pinchang et al., 2012; Wang Wu, 2012), Newcastle disease (Wu Changxin et al., 1999), intestinal parasitic diseases (Hu Yan, Hu Hui, 2002; Foronda et al., 2004; Liu Yunlong et al., 2011; Huang Chao et al., 2015, 2017) and other diseases and their treatment measures. In recent years, there have been many reviews conducted on the threat factors and conservation management of green peafowl as well as reports of the conservation level, endangered status, threats and conservation, and management measures that have been implemented (Wen Xianji et al., 1995; Yang Xiaojun et al., 1997; Xie Yichang, 2016; Kong Dejun and Yang Xiaojun, 2017; Fu Changjian et al., 2019; Gu Bojian and Chen Yuqian et al., 2019; Li Binqiang et al., 2018).

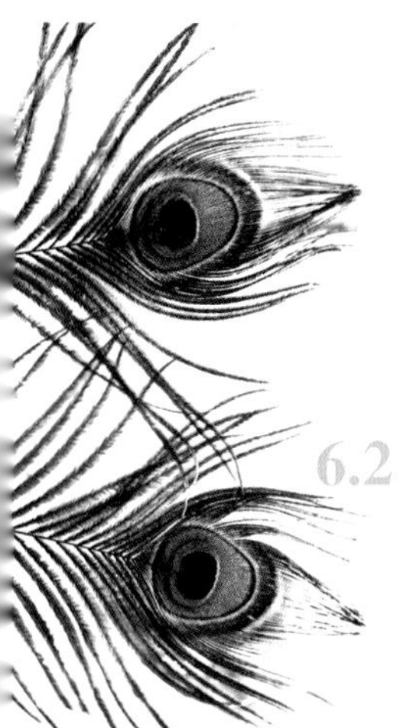

6.2 Current Research of Wild Green Peafowl in China

Liu Zhao et al. investigated the foraging habitat of green peafowl in its distribution area of the Shiyang River Valley of the upper reaches of Yuanjiang River in Yunnan from March to April and from October to November in 2007 and identified 21 key ecological factors (Fig.6–3). A detailed investigation has been made on the choice of the green peafowl's habitat, their choice of foraging sites in different seasons, and the influence of human disturbance on green peafowl decision making when selecting feeding spots.

In 2015, Wen Yunyan, et al. conducted surveys of green peafowl and performed the required monitoring in the concentrated distribution area of green peafowl in Konglonghe Prefectural Nature Reserve of Shuangbai in 2015, using plotting combined with IR triggered automatic cameras. They obtained a large number of photos and related videos of green peafowl in the wild so that the time-table of courtship, mating, and the incubation of green peafowl was initially determined, and problems present in the conservation of green peafowl were analyzed (Fig.6–4).

▲ Fig.6–3 Distribution area of green peafowl of the Shiyang River Valley of the upper reaches of Yuanjiang River (Photo by the courtesy of Wang Fang)

▲ Fig.6–4 Distribution area of the green peafowl of the Konglonghe river basin (Photo by the courtesy of Wang Fang)

Using green peafowl of Mojiawan in Konglonghe Natural Reserve of Chuxiong, Yunnan Province as a comparison, Li Xu, et al. analyzed the differences between various environmental characteristics and their comprehensive influence on the choice of green peafowl's habitat in 2016. The results made it clear that food and shelter conditions were the main factors affecting the choice of green peafowl foraging selection (Fig.6–5).

▲ Fig.6–5 Konglonghe Natural Reserve of Chuxiong, Yunnan Province (Photo by the courtesy of Wang Fang)

Kong Dejun and others summarized the basic biological data of green peafowl in 2017, including morphological characteristics, subspecies differentiation, habitat and reproductive characteristics, feeding habits, population size, and distribution. They also pointed out the problems regarding green peafowl conservation in China (Fig.6–6).

▲ Fig.6–6 Wild green peafowl foraging in the forest (Photo by the courtesy ofthe green peafowl research team of Yunnan University using IR camera)

Shan Pengfei et al. (2018) first discovered the nest of wild green peafowl and described their habitat. At the same time, the weight and size of the green peafowl eggs were measured, and were consistent with previous descriptions by Zheng Guangmei and Yang Lan et al.

Wang Fang et al. (2018) adopted IR cameras to monitor wild green peafowl in Xinping County, Yunnan Province. The results showed that there were wild green peafowl in Xinping County, and they made an initial determination about their distribution and habitat range (Fig.6–7–Fig.6–9).

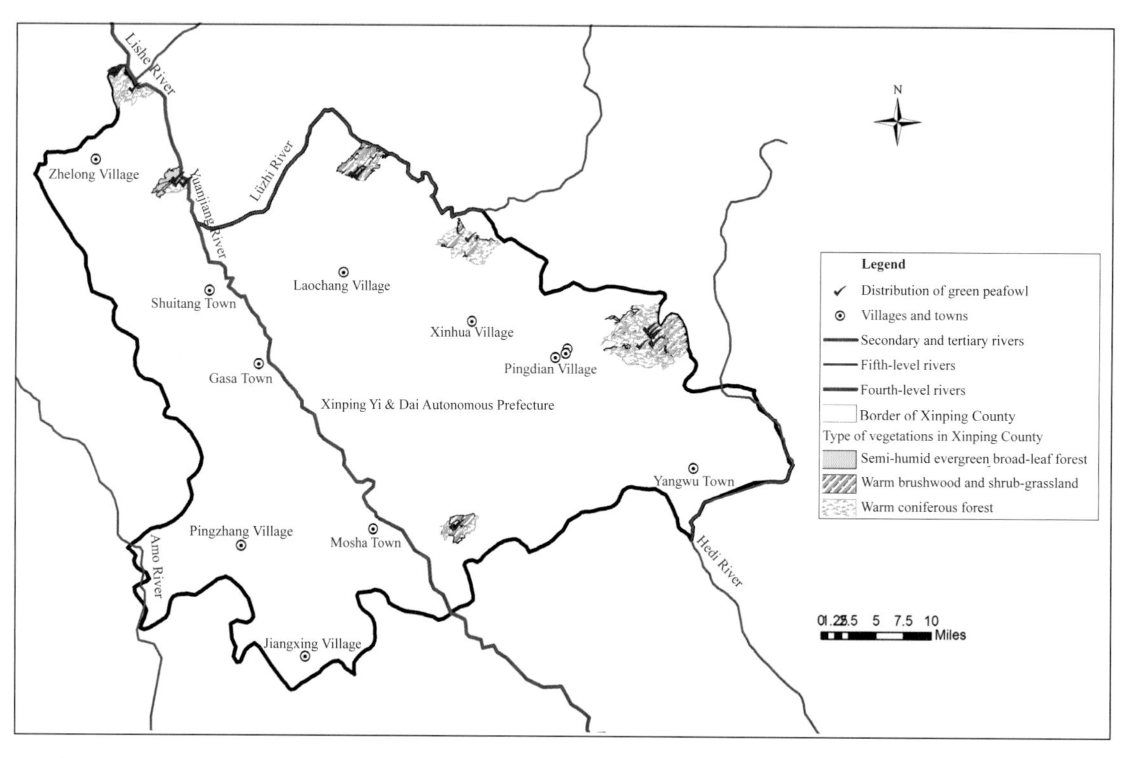

▲ Fig.6–7 A map of the distribution status of green peafowl in Xinping County completed by the green peafowl research team of Yunnan University (Photo by the courtesy of Chen Mingyong)

▲ Fig.6–8 The research team of Yunnan University found and measured the green peafowl's foot traces on the sample line (Photo by the courtesy of Chen Mingyong)

▲ Fig.6–9 The green peafowl research team of Yunnan University is placing IR cameras in the field. (Photo by the courtesy of Chen Mingyong)

Wang Fang et al. (2018) obtained tens of thousands of images of wild green peafowl through continuous monitoring by IR cameras throughout a whole year. The statistics showed that the population of wild green peafowl in Xinping County was somewhere between 127 and 151 at that time (Fig.6–10).

▲ Fig.6–10 An IR camera used to monitor green peafowl (Photo by the courtesy of Wang Fang)

Wang Fang et al. (2018) analyzed the diversity of birds and mammals associated with wild green peafowl and introduced the 2 × 2 contingency table to analyze the interspecific relationship between green peafowl and those birds and mammals (Fig.6–11).

▲ Fig.6–11 The green peafowl's food competitor — Gallus gallus (Photo by IR cameras)

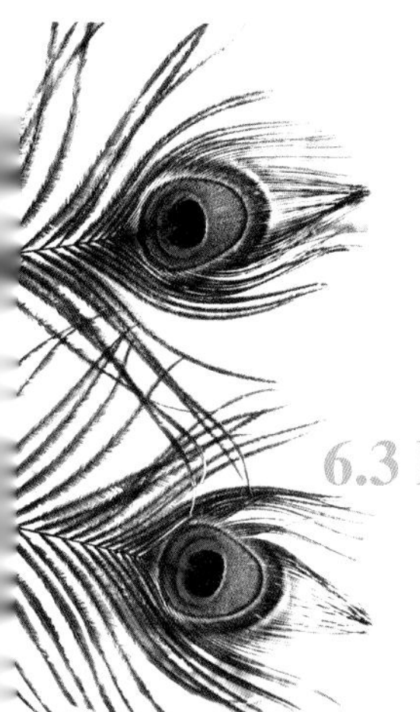

6.3 Major Problems Facing Wild Green Peafowl in China

Yang Xiaojun et al. (2017) believed that the main risk factors for wild green peafowl in China included death (poisoning, poaching), habitat loss (deforestation, mining, hydropower expansion, highway construction, and deforestation), environmental disturbance (village, grazing, picking), and conservation management (mostly distributed outside the conservation area). Furthermore, the current population status of green peafowl as consisting of small isolated populations, generally smaller groups, habitat degradating (Fig.6–12) and increasing human interference has dramatically pushed the species towards extinction.

Hua Rong et al. (2018) deemed that the main threats facing the green peafowl in China were: ① rampant poaching; ② habitat destruction caused by deforestation, land reclamation, road construction, and construction of hydropower stations; ③ habitat fragmentation leading to isolated distribution of small populations and inbreeding; ④ pesticide-coated seeds, rodenticides, poultry infectious diseases and infectious diseases within the population; ⑤ the escape of captive blue peafowl from the local villagers' cages, causing genetic pollution of wild green peafowl.

▲ Fig.6-12 Habitat degradation (Photo by the courtesy of Wang Fang)

6.3.1 Severe Habitat Loss

Firstly, the distribution range of the peafowl has declined sharply. Due to alternative planting, engineering construction, deforestation, mining, sand fetching from riverbeds, and invasion of other species, the green peafowl's habitat has shrunk drawatically. Compared with the 1990s, the distribution range of green peafowl has dropped from 42 counties to 19 counties. Secondly, habitat fragmentation is severe. Green peafowl populations in Yunnan Province are distributed in fragments. In the middle and upper reaches of Yuanjiang River where the population density is high, green peafowl are also distributed into fragmentary, small groups due to the construction of roads, power stations, farmland, villages, economic forests, etc. and the obstruction of the natural habitats that they incur. Because of this, it is difficult for green peafowl to carry out efficient gene exchanges. At present, the Dawan Power Station and the first grade Power Station in Xiaojiang River have been built in the middle and upper reaches of Yuanjiang River, which are important habitats for the green peafowl. Although the construction of the first grade Power Station in Jiasa River and the second grade Power Station in Xiaojiang River have been suspended, the valley areas below 680 meters will be submerged if the construction was resumed. The river beaches in this area are currently important places for sand baths and serve as courtship sites for green peafowl. If water is stored there, the green peafowl's suitable habitat will be submerged. In addition, the Dawan Power Station is connected end to end with the first grade Power Station in Jiasa River. As the river water (reservoir area) widens, the green peafowl population on the east and west banks are isolated, which seriously affects the gene exchange between the populations. Thirdly, the suitability of habitat has declined. Due to the transformation of agricultural production methods, the use of farming machines, lawn mowers and other machinery and pesticides have generated noise and pollution, which has had a negative impact on green peafowl; activities of community residents such as picking and grazing under the forest have interfered with the reproduction of green peafowl.

6.3.2 Weak Conservation Foundation

Firstly, management mechanisms surrounding green peafowl are imperfect. Green peafowl are mainly distributed along the Red River, Lancang River and Nujiang River. Watershed management involves water conservancy, land, forestry, agriculture, and other departments, so that management authority is decentralized. There are conflicts of interest between departments, and a lack of communication and coordination between departments. In addition, there is no unified watershed management concept. All of these factors contribute to the destruction and degradation of the green peafowl habitat. Secondly, conservation and management capabilities are weak. Green peafowl are distributed more extensively in Shuangbai County and Xinping County, yet there is only one reserve there, namely the Konglonghe Prefectural Nature Reserve. Due to relatively inefficient protection, lack of capital investment, backward infrastructure and equipment, insufficient staff, delayed publicity and education, etc., the conservation and management in the reserve is not enough. Thirdly, the conservation doesn't cover a large area. Among the various levels of nature reserves established in Yunnan Province, 7 reserves still have green peafowl with a population of 170 to 183, but nearly two-thirds of the green peafowl populations are outside the nature reserves. Therefore, the scope of the conservation area should be further expanded.

6.3.3 Weak Foundation for Scientific Research and Monitoring

Firstly, the systematic and basic research is week. State and local funding for research on green peafowl is relativcly small, and there is a shortage of research data on the use and selection of habitat, reproduction, and the population's genetic structure. Therefore, it is difficult to propose targeted conservation and management measures based on the ecological and biological characteristics of green peafowl. Secondly, the monitoring system is imperfect. Green peafowl monitoring is only carried out in a partial area of the Konglonghe Prefectural Nature Reserve. A complete monitoring system has not been established in the province. Monitoring mainly relies on human patrols. The lack of modern monitoring equipment and methods makes the monitoring capabilities hard to carry out to a properly effective degree.

6.3.4 Publicity and Education Lagging Behind

Firstly, publicity and education are insufficient. As publicity and education of green peafowl conservation is rarely carried out, it is difficult for the public to identify the exotic species of blue peafowl and green peafowl. Individual media and businesses confuse the two when using the image of peafowl to promote products. Therefore, the public knows little of the endangered degree of green peafowl, the level of conservation, and the risks and difficulties faced by conservation; green peafowl are mostly common in remote and poor mountainous areas, but the awareness of community conservation is weak there. Secondly, the guidance of ecological agriculture is not enough. With the changes in agricultural production methods, farmers sow pesticide-containing "coated seeds" or crop seeds soaked in pesticides, spray pesticides in the fields, scatter poison for rodents, etc., of which the green peafowl may unknowingly eat and die.

6.3.5 Unstable Capital Investment

Since 2009, the forestry department has only invested more than 3 million yuan. In 2017, the State Forestry Administration of the People's Republic of China issued a special conservation fund of 1.65 million yuan. In 2017 and 2018, Yunnan Forestry Department carried out green peafowl population and habitat surveys in the upper reaches of the Yuanjiang River. The appraisal fund was 195 thousand yuan, and the preparation fund for the Yunnan green peafowl conservation plan was 100,000 yuan. Due to the lack of nationally allocated funds for wildlife conservation, and the lack of special funds for conservation in Yunnan Province, it is difficult to effectively carry out work concerning the management, monitoring, and public education of green peafowl habitats.

According to analysis, wild green peafowl in Xinping County, Yuxi City, Yunnan Province is under continuous threats because of habitat loss and population decrease. Thus, it is urgent to rescue and protect the green peafowl. The primary reasons for the changes in the distribution of wild green peafowl in China are grazing, large-scale project construction, poaching, habitat degradation, human disturbance, and other external factors as well as germplasm factors of green peafowl itself (Fig.6–13–Fig.6–15).

If we analyze each type of threat factor and assign values for the three aspects including scope, severity, and irreversibility, the highest threat is the habitat damage, followed by grazing, graveling, mining, hunting, etc. Other threats include logging, building roads, farming, gathering, forest fires, etc.

▲ Fig.6–13 A large number of cultivated plantations around the green peafowl distribution area (Photo by the courtesy of Wang Fang)

▲ Fig.6–14　Plantation after development in the green peafowl distribution area (Photo by the courtesy of Wang Fang)

▲ Fig.6–15　Engineering construction site in the green peafowl distribution area (Photo by the courtesy of Wang Fang)

With the development of the economy, the construction of large-scale engineering projects such as hydropower stations and power plants have also severely damaged the habitat of green peafowl and caused geographical isolation among green peafowl populations. In addition, human disturbances such as planting, illegal poaching, logging, etc. have caused the population of green peafowl and the range of their habitat distribution to shrink.

▲ Fig.6–16 Large amount of logging in the green peafowl distribution area (Photo by the courtesy of Wang Fang)

The high amount of logging every year not only leads to the habitat loss, but also causes great disturbance to the survival of green peafowl (Fig.6-17).

▲ Fig.6-17 Traces of manual cutting of pine resin in the green peafowl habitat (Photo by the courtesy of Wang Fang)

A large number of sand mining boats in the Red River basin continue to carry out sand mining and sand washing, making the water turbid, and emitting noise pollution which affect the lives of local peafowl (Fig.6-18).

▲ Fig.6-18 There are many sand mining boats in the river in the green peafowl habitat (Photo by the courtesy of Wang Fang)

6.4 Current Conservation Status of Wild Green Peafowl in China

The conservation of wild green peafowl in China has not yet been agreed upon different levels of government and departments of the distribution areas. Not every nature reserve is equipped with professionals, and management personnel fail to have a comprehensive and scientific understanding of green peafowl. The monitoring equipment of the conservation bureaus in the management areas is incomplete or outdated, making it difficult to conduct research activities on green peafowl quickly and comprehensively. At present, the conservation of wild green peafowl in China has received extensive attention from all walks of life. In order to better protect the Chinese wild green peafowl, Yunnan Province has made a list of conserved species with extremely small populations and focuses on its conservation and rescue activities. The Kunming Animal Research Institute of the Chinese Academy of Sciences carried out a survey of the distribution of wild green peafowl across Yunnan Province. The Alashan League and Yunnan Forestry and Grassland Administration have established green peafowl conservation communities in Xinping County, Yuxi City, Yunnan Province, and concentrated their efforts on the investigation and some universities like Yunnan University and Southwest Forestry University have began to concentrate scientific ornithology research on the investigation and conservation of wild green peafowl (Fig.6–19–Fig.6–21).

▲ Fig.6–19 Conducting field investigation of green peafowl (Photo by the courtesy of Wang Fang)

▲ Fig.6–20 Billboard of the green peafowl conservation (Photo by the courtesy of Wang Fang)

▲ Fig.6–21 Symposium on the conservation of green peafowl (Photo by the courtesy of Cao Shun)

7 Peafowl Culture

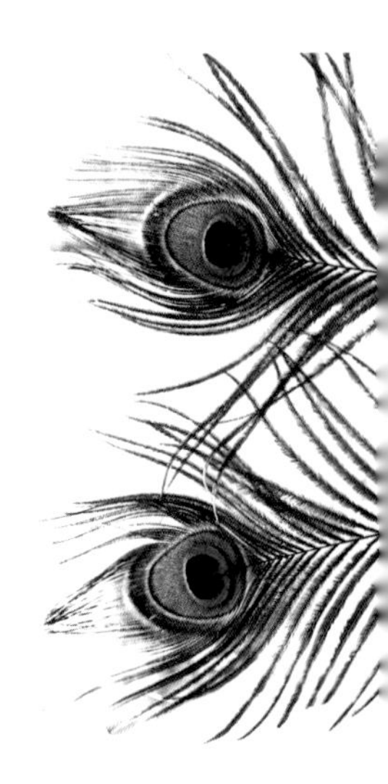

7.1 Cultural Values

Since ancient times, green peafowl have a close bond with mankind throughout history, green peafowl have become especially intertwined with the cultures of ethnic minorities in China. Green peafowl are also well-known in fields like literature, mythology, religion and national art (Fig.7–1).

▲ Fig.7–1 Peafowl *and Red Plum Blossom*, collection of the Palace Museum

A Pair of Peacocks Southeast Fly, the best-known poem from *Yuefu Poems of Han Dynasty*, tells us a story, "During the reign of Jian'an (196–219) in the Eastern Han Dynasty there was a local official in the prefecture of Lujiang called Jiao Zhongqing, whose wife, Liu Lanzhi, was sent away by his mother and vowed never to marry again. Compelled by her family to break her vow, she had no recourse but to drown herself in a pool. On hearing the news, Zhongqing hanged himself on a tree in the courtyard." The poem begins with "A pair of peacocks southeast fly; at each mile they look back and cry." It continues at another part: "Their families, after they died; buried them by the mountainside." *

In Chinese myths and religious legends, the peafowl is considered the king of birds, since it is the child of Phoenix and the sibling of Dapeng Golden Winged Bird. Later, the peafowl was granted the title of Peafowl Daming King Bodhisattva by the Buddha, so it is also called Peafowl Daming King. (Wang Yanbo, 2018)

In Yunnan Province, many ethnic minorities are very fond of peafowl. For example, the Yi nationality in Central Yunnan and the Dai nationality in Southern Yunnan worship peafowl as their significant totem. Dai people believe that peafowl are kind and intelligent, love freedom and peace the most, and symbolize auspiciousness and happiness.

Spanning a history of more than 1,300 years, the Dai people live a quiet farming life due to the humid and hot climate in the subtropical zone. Most of them believe in Hinayana Buddhism and pursue the realm of wisdom and emptiness, which more closely resembles the peafowl's nimbleness and gentleness.

(* Here, the translation of the poem *A Pair of Peacocks Southeast Fly* quotes from the famous translator Xu Yuanchong's *Illustrated Poems of the Han, Wei and Six Dynasties*.)

7.2 The Peafowl and the Phoenix

The Phoenix has a variety of archetypes, such as the Chinese copper pheasant, the peafowl, the eagle, the swan, and the swallow. As archetypes of the Phoenix, peafowl are an auspicious bird worshipped by the Dai people and common in the area where they live. According to historical records, "Peafowl nest in some [of their] courtyard trees." It is apparent that peafowl and the Dai people are tightly bound to each other. The Dai Autonomous Prefecture of Xishuangbanna and the Dehong Autonomous Prefecture are both known as the "Hometown of Peafowl." The Dai people often regard peafowl as a symbol of their ethnic spirit, and they use the peafowl dance to express their wishes and ideals, and to praise the wondrous aspects of their lives. The peafowl dance is a favorite ancient folk dance of the Dai people, and is popular in the entirety of the Dai nationality area. The most wonderful types of peafowl dance include those performed in areas like Ruili, Gengma, Mengding and Mengla. In the past, performers often wore masks and imitated peafowl's movements — walking in the forest, drinking water, splashing in the water, chasing, playing games, shaking wings, displaying tail feathers, rotating, etc.

The peafowl is the most intelligent animal in Dai literature. In the folktale *A Story of Two Princes*, two brothers, who were princes, were denounced by a wild male goose for shooting a female wild goose with an arrow. The king sentenced the two princes to death. At the request of the queen, the executors released the two princes, who then wandered in a foreign country and shot the king's peafowl there. After the eldest prince ate the head of the peafowl, he glowed like sunshine and became a king. When the younger prince ate the liver of the peafowl, gold and silver began to come out of his mouth as he spoke... Nowadays, the Dai people take peafowl as mascots and perform the peafowl dance. Peafowl are an important theme in brocade, embroidery, paper cutting and painting. People also incorporate peafowl patterns into their house decoration (Fig.7–2–Fig.7–3) and women's jewelry. All of these are closely related to the bird totem worship of the ancient Yue people, the ancestors of the Dai nationality. The Dai people like to have peafowl as tattoo pattern, a clear indicator of their hope to be blessed by the peafowl, the archetype of the Phoenix, so that they may enjoy happiness, peace and good luck. The peafowl was embroidered on the military flag of the hero "Sangluo" in the heroic epic *Lifeng* of the Dai nationality. Phoenix patterns were also found on the flag of Xishuangbanna governor's honor guard. Peafowl or Phoenix was the emblem of the group regime, which had a close correlation with the early bird totem worship of the Dai people (Dao Chenghua, 2009).

▲ Fig.7–2 Peafowl sculpture from Burmese Temple in Dai Village (Photo by the courtesy of Tang Yongjing)

▲ Fig.7-3 Sculptures of green peafowl, white peafowl, and a white elephant from Burmese Temple in Dai Village (Photo by the courtesy of Tang Yongjing)

7.3 Peafowl Dance

The peafowl dance has a long history in Dai nationality area. The ancient murals and carvings of Burmese temple in Dai village display images of the peafowl with a vivid human face and bird body, bearing much similarity to the actual peafowl dance of today which utilizes peafowl inspired masks and clothing in the Dai village, pointing distinctly towards the long history of the peafowl dance in the area. The Peafowl dance is called *Galuoyong*, *Fanluoyong*, or *Gananluo* in the Dai language. Due to their strong belief in Hinayana Buddhism, Dai people dislike violent sports. This is why the peafowl dance has become the most popular dance there. The Dai town is known as the "Hometown of Peafowl." In the past, peafowl with beautiful postures danced in the illuminated morning light and in the glow of the sunset. The *Unofficial History of Nanzhao* in Ming Dynasty once recorded that "people, old or young, dance in the wedding; they play the Lu Sheng for the peafowl dance."

People offer peafowl feathers to the Buddha and perform peafowl dance for good luck. The Peafowl dance has naturally become the soul of Dai dance culture and constitute an important part of their aesthetic views (Fig.7–4–Fig.7–7).

▲ Fig.7–4 Dai people's peafowl dance (Photo by the courtesy of Hasegawa Kiyoshi)

▲ Fig.7–5 Dai people's white peafowl dance (Photo by the courtesy of Hasegawa Kiyoshi)

▲ Fig.7–6 The peafowl costume parade during the Water Splashing Festival of Dai Nationality in Xishuangbanna (Photo by the courtesy of Chen Mingyong)

▲ Fig.7–7 Local people's peafowl costume in Cambodia (Photo by the courtesy of Chen Mingyong)

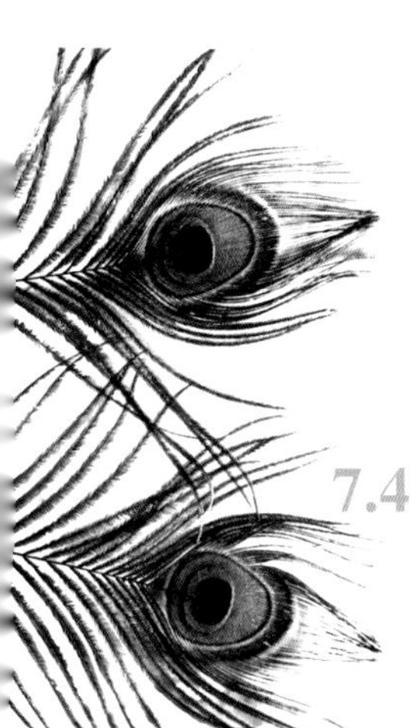

7.4 Research on the Name of "Peafowl" in Xinjiang and the Origin of the Name "Peafowl River"

Wang Shouchun (2015) conducted a topical research on the origins of the name of peafowl in Xinjiang and the origin of the name of "Peafowl River." After his research, he believed that the "Peafowl" in Qiuci (an ancient state), recorded in *The History of Wei Dynasty: Biography of Western Region*, was not the real peafowl, but a bird called "Kum-tuche" (translated as "shatuti" in Chinese) by local residents. Kum means "sand" in Uyghur language (a semantic translation), and "Tuti" is the transliteration of "tuche." Therefore, "Shatuti" is actually a mixed translation of "Kum-tuche." Its exact semantic translation into Chinese should be "sand sparrow" or "sparrow on the sand," indicating that the bird mainly lives in a sandy or desert environment. In modern bird taxonomy, several species of birds belong to the Passeriformes, Corvidae family (Crow was mistaken for "duck" in the original text.), which include Podoces hendersoni Hume, Podoces humilis Hume and Podoces biddulphi and Pyrrhocorax pyrrhocorax (These crows were mistaken for ducks in the original text). It was easily mistaken because the habits and ecological characteristics of "Kum-tuche" were very similar to "Peafowl" recorded in *The History of Wei Dynasty*. Furthermore, the "Peafowl" described in *Traveling in Xinjiang* was probably the Pyrrhocorax pyrrhocorax (mistaken for "duck" in the original text) because the first three kinds of crows mentioned above mainly live on hills or in the Gobi desert. Only the Pyrrhocorax pyrrhocorax lives atop high mountains. Historically, the bird called "Kum-tuche" by people of Tarim Basin was closely related to local folk customs. *The History of Wei Dynasty* recorded that Qiuci people ate "Kum-tuche" and Uyghur women used to decorate their hats with feathers

▲ Fig.7–12 Large peafowl sculpture (Photo by the courtesy of Chen Mingyong in Ancient Dian State)

of Podoces hendersoni Hume because of their beautiful purple-blue color or purple-black metallic luster. The historical record of "Peafowl" in Tarim Basin could not be the green peafowl, but the "Kum-tuche" in Xinjiang (Podoces hendersoni Hume). The Peafowl River (transliterated as Kunqidaliya in Uyghur) is an important river in Xinjiang. Rising in Bosten Lake, its water supplies come from Kaidu River. After being naturally regulated through the Bosten Lake, it flows out from the southwest corner of the lake and passes through the Weihu Lake to converge into the Peafowl River. The name "Peafowl River" originally referred to the lower reaches past Korla. The name of the Peafowl River in the mid-Qing Dynasty was Kaidu River or Haidu River, which was named according to the features of the river. From the convention of naming places in Xinjiang, many place names or river names are based on natural characteristics. For example, "Kaidu River" is named because of the bend formed by the Peafowl River in the west of Korla. And others are named after animals or plants. For example, the name of Butahaizi (a lake in Xinjiang) derives from the habitat of waterfowls all year round. "Peafowl River" is possibly named after "Kum-tuche" (Podoces hendersoni Hume).

8

An Outlook on the Research and Protection of Wild Green Peafowl

8.1 Strengthening Basic Research of Green Peafowl

Detailed research results on green peafowl biology, ecology, and conservation biology are the prerequisite and basis for developing scientific conservation policies and strategies while implementing conservation management measures. Given that the basic research on wild green peafowl in China is still weak, it is recommended to organize relevant scientific research units and scientific researchers from colleges and universities to conduct regular surveys on the distribution and number of green peafowl populations throughout the region, research on population activity patterns and population dispersal, research on food habits and behavior, research on habitat selection, research on breeding ecology, potential habitat prediction, research on key habitat restoration techniques, and research on the impact of human interference on green peafowl alongside their conservation biology, genetic diversity, epidemic focus, diseases, etc. Research will lay a solid foundation for the basic ecological data of green peafowl and provide scientific theories for formulating relevant management policies and protection measures.

8.2 Strengthening Conservation Management of Green Peafowl Population

Expand the area of existing nature reserves in areas where green peafowl populations are concentrated, strengthen nature reserve institutions and capacity building, develop nature reserve protection and management plans, specialized protection plans for green peafowl, and increase patrol and monitoring efforts and law enforcement.

It should be strictly prohibited to hunt, capture, or harm wild green peafowl. Efforts should be made to establish public forest security, enhance law enforcement, promote forestation and the nurturing of grassland, and facilitate the construction of nature reserves to carry out proper law enforcement there. Measures should be taken to combat illegal poaching of green peafowl and other wildlife. Through protection, monitoring, artificial multiplication, science education, and other measures, the steady population growth of green peafowl must be ensured. We will also make sure that the habitat of green peafowl is effectively preserved through protection and restoration. We can effectively improve social participation and public awareness by encouraging social forces to participate in the protection and pumping more efforts into publicity.

8.3 Strengthening the Management and Control of Green Peafowl Habitats

Present and build green peafowl protection corridors between existing green peafowl population distribution areas, investigate and evaluate the quality of habitats in existing distribution areas, develop habitat protection and restoration programs based on the characteristics of each area, and first incorporate habitat protection into government work plans and financial budgets.

We will plan and construct special national parks, nature reserves, and protection areas for green peafowl to strengthen the protection of their original habitats. The impact evaluation criteria should be formulated, and the examination and approval procedures should be strict for the construction of large bases within the habitats of green peafowl. Human activities such as stone quarrying, sand mining, firewood gathering, and land cultivation within the scope of potential habitats and ecological corridors should be prohibited. Blasting activities such as mining and quarrying around potential habitats and ecological corridors should be halted. Human activities such as camping and crossing within the scope of potential habitats and ecological corridors should be prevented. Green peafowl's habitats can be protected and managed by patrols to prevent human destruction and forest fires. At the same time, the habitats can be monitored by men and IR cameras.

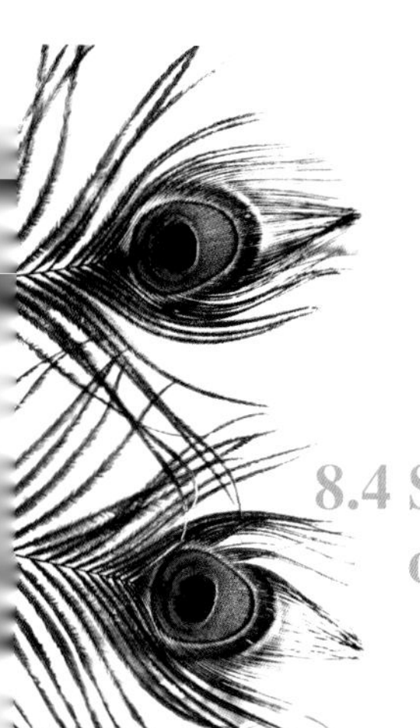

8.4 Strengthening Publicity and Education on the Protection of Wild Green Peafowl

Construct a protection and propaganda system for the green peafowl and its habitat, such as building a nature museum and a cultural museum to carry out the protection and promotion of the green peafowl and its habitat in the form of exhibitions of nature, education, science and promotion activities.We will organize experts to compile publicity manuals and other materials for the protection of green peafowl, and educate students in universities, middle schools, primary schools and kindergartens to enhance their awareness of protection. Using platforms such as WeChat and microblogs, public protection awareness advertisements through films and videos; going into campuses and communities to promote green peafowl protection and knowledge of laws and regulations related to nature protection through distribution of propaganda materials, special lectures, home visits, etc., to enhance the involvement of all sectors of society and the public, especially residents in the distribution areas of green peafowl who love green peafowl, protect them, and raise the protection awareness of society.

8.5 Carrying out Artificial Breeding and Reintroduction of Green Peafowl

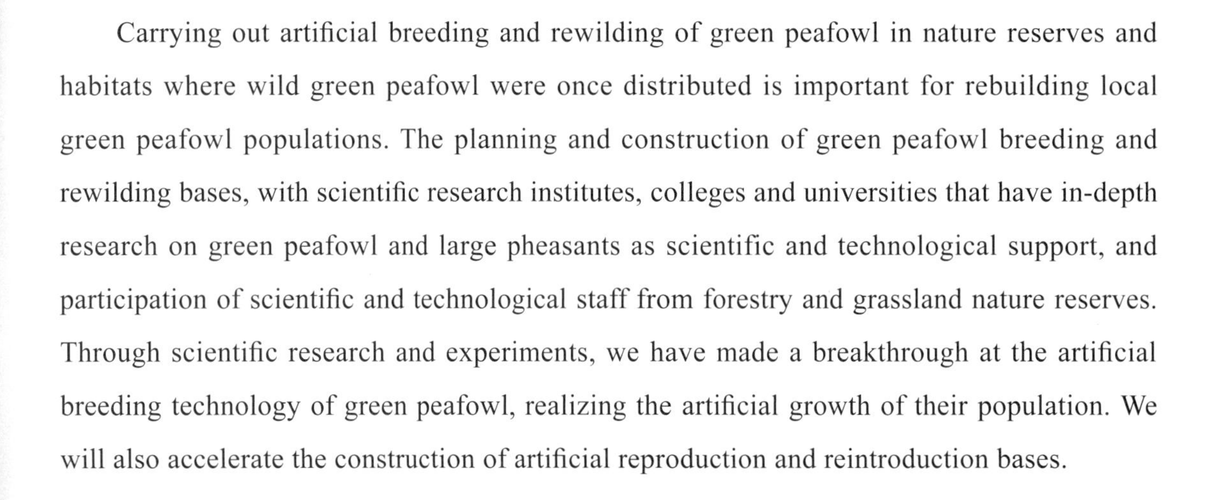

Carrying out artificial breeding and rewilding of green peafowl in nature reserves and habitats where wild green peafowl were once distributed is important for rebuilding local green peafowl populations. The planning and construction of green peafowl breeding and rewilding bases, with scientific research institutes, colleges and universities that have in-depth research on green peafowl and large pheasants as scientific and technological support, and participation of scientific and technological staff from forestry and grassland nature reserves. Through scientific research and experiments, we have made a breakthrough at the artificial breeding technology of green peafowl, realizing the artificial growth of their population. We will also accelerate the construction of artificial reproduction and reintroduction bases.

Reference

[1] Ai HuaiSen. Diversity and conservation of pheasants in Gaoligong Mountain[J]. *Zoological Research*, 2006, 27 (4):427–432.

[2] Bai Na. Explore the geographical distribution of peacocks in Chinese historical period from historical materials[J]. *Journal of Wenshan College*, 2015,28(4):56–58.

[3] Bao Wenbing, Chen Guohong, Shu Jingting, et al. Screening of primers and genetic diversity analysis of microsatellite in peafowls[J]. *Heredity*, 2006, 28(10): 1242–1246.

[4] Chen Jing. Diagnosis and treatment of coccidiosis in green peafowl[J]. *Aquaculture Technical Consultant*, 2011,(11):217.

[5] Chang Hong, Ke Yayong, Su Yingjuan, et al. Study on random amplified polymorphic DNA of wild and caged green peafowl populations[J]. Heredity, 2002,24(3):271–274.

[6] Cong Peihao, Zheng Guangmei. Study on hatching and brood behavior of *Tragopan temminckii*[J]. *Journal of Beijing Normal University (Natural Science)*, 2008,44(4): 405–410.

[7] Cui Peng, Kang Mingjiang, Deng Wenhong. The selection of foraging habitat for the sympatric tragopan and blood pheasant in the breeding season[J]. *Biodiversity*, 2008,16(2): 143–149.

[8] Dang Xinyan, Ma Guoqiang, Tang Yangchun, et al. Analysis on the nutritional status of blue peafowl in zoo[J]. *Journal of Southwest Forestry University*, 2014,34(1):102–105.

[9] Duan Yubao, Li Yuan, Chen Xi, et al. Analysis of genetic differences between blue peafowl and green peafowl based on mitochondrial gene[J]. *Genomics and Applied Biology*, 2020,39(2): 547–552.

[10] Feng Feng, Bi Yantai, He Xiangbao. Observation on breeding behavior of domestic green peafowl[J]. *Chinese Journal of Wildlife*, 2007, 28(4):16–17.

[11] Fu Changjian, Qiu Huanlu, Yu Jia. Endangered status and its protection of green peafowl in China[J]. *Chinese Journal of Wildlife*, 2019,40(1):233–239.

[12] Guo Baoyong, Wang Zhisheng, Lan Daoying. Current situation of wildlife resources in Nangunhe Nature Reserve [J]. *Chinese Journal of Wildlife*, 1999,20(4):46–47.

[13] Guo Dongsheng, Zhang Zhengwang. Bird Ecology in China[M]. Chongqing: Chongqing University Press, 2015.

[14] Wang Jiao, Zhang Chongliang, Xue Ying, et al. Interspecies association of main fishes in Haizhou Bay and its adjacent waters[J]. *Chinese Journal of Applied Ecology*, 2020,31(1):293–300.

[15] He Yeheng. Historical changes of rare birds in China[M]. Changsha: Hunan Science and Technology Press, 1994.

[16] He Ping, Lu Yuyan. Histological observation of the digestive system of green peafowl[J]. *Journal of Shaanxi Normal University* (Natural Science Edition), 2002,30(4):92–95.

[17] Huang Chao, Lu Yongchao, Yang Juan, et al. Investigation on intestinal parasites of green peafowl in Yuantongshan Zoo[J]. Animal Husbandry and Feed Science, 2015,36(10):8–9.

[18] Huang Chao, Liu Anrong, Yang Juan, et al. Intestinal parasitic characteristics of green peafowl in zoo [J]. *Chinese Journal of Veterinary Medicine*, 2017,53(1):53–55.

[19] Hua Rong, Cui Duoying, Liu Jia, et al. Investigation on the population status of green peafowl in China[J]. *Chinese Journal of Wildlife*, 2018,39 (3): 681–684.

[20] Jia Lanpo, Zhang Zhenbiao. The fauna of Xiawanggang site in Xi county, Henan Province [J]. *Cultural relics*, 1977(6):41–49.

[21] Jiao Yuanmei, Liu Xin, Li Rong, et al. Study on functional zoning of Ailao Mountain National Park based on potential habitat of green peafowl[J]. *Tourism Science*, 2017,31(3):75–84.

[22] Kuang Bangyu. Peafowl in southern Yunnan[J]. *Biology Bulletin*, 1963,11(4):17–18.

[23] Kong Dejun, Yang Xiaojun. Green peafowl and its conservation status in China[J]. *Biology Bulletin*, 2017,52(1):9–11,64.

[24] Lan Yong. Chinese Historical Geography[M]. Beijing: Higher Education Press, 2002.

[25] Li Binqiang, Li Pengying, Yang Jiawei, et al. An infrared camera survey of bird and animal

diversity in Qinghua Green Peafowl Nature Reserve of Weishan Mountain, Yunnan Province[J]. *Biodiversity*, 2018, 26 (12): 1343–1347.

[26] Li Jian, Wang Liping, Wang Wen. Histological observation of digestive system of green peafowl at 2 months of age[J].*Journal of Northeast Forestry University*, 2004,32(2):62–64.

[27] Li Mingfu, Li Sheng, Wang Dajun, et al. A study on the diurnal activity patterns of the winterspring twitterhorn in Tangjiahe Nature Reserve, Sichuan Province[J]. *Sichuan Animal*, 2011,30(6): 850–855.

[28] Li Xu, Liu Zhao, Zhou Wei, et al. Habitat selection and spatial distribution of green peafowl in spring in Konglonghe Nature Reserve, Chuxiong, Yunnan[J]. Journal of Nanjing Forestry University, 2016,40(3):87–93.

[29] Li Ruinian, Lin Haiyan. A preliminary study on the distribution of green peacocks along the highway construction project from Ruili to Menglian[J]. Proceedings of the Science and Technology Annual Conference of the Chinese Society for Environmental Science, 2018:493–497.

[30] Liu Fang, Li Linna, He Baojie, et al. Breeding and reproduction of green peafowl[J]. *Special Economic Animals and Plants*, 2012,3:12–13.

[31] Liu Xiao-bin, Wei Wei, Zheng Xiaoguang, et al. *Journal of Zoology*, 2017,52(2):194–202.

[32] Liu Zhao, Zhou Wei, Zhang Rengong, et al. The selection of foraging sites for green peafowls in different seasons in the Shiyangjiang River valley in the upper reaches of Yuanjiang River, Yunnan Province[J]. *Biodiversity*, 2008,16(6):539–546.

[33] Lu Taichun. Rare and endangered wild chickens in China[M]. Fuzhou: Fujian Science and Technology Press, 1991.

[34] Luo Aidong, Dong Yonghua. Investigation on the population size and distribution of wild green peafowl in Xishuangbanna[J]. *Chinese Journal of Ecology*, 1998, 17(5):6–10.

[35] Ma Jianzhang. Practical handbook of wildlife protection in China[M]. Beijing: Scientific and Technical Documents Press, 2002.

[36] Ouyang Yina, Yang Zeyu, Li Dalin, et al. Genetic differentiation of green peafowl and blue peafowl using cytochrome b gene[J].*Journal of Yunnan Agricultural University*, 2009,24(2):220–224.

[37] Song Zhiyong, Yang Hongpei, Yu Dongli, et al. Population status of the grey peafowl's pheasant in Xishuangbanna[J]. *Journal of West China Forestry Sciences*, 2018,7(6):67–72.

[38] Shan Pengfei, Wu Fei, He Jiafei. The green peafowl rarely seen[J]. *Forest and Human*, 2017(2):62–63.

[39] Tang Songyuan, Li Li, Duan Wenwu, et al. Artificial breeding technology of blue peafowl[J]. *Hunan Forestry Science and Technology*, 2019,46(3):75–79.

[40] Wang Yanbo. Research on the distribution change and culture of peafowl in historical period in China[D]. Northwest A & F University, 2018.

[41] Wen Huanran and He Yecheng. The peafowl in ancient China[J]. *Fossils*, 1980(3):8–9.

[42] Wen Huanran and He Yecheng. A Study on the changes of plants and animals in Chinese historical period[M]. Chongqing: Chongqing Press, 1995.

[43] Wen Huanran, He Yecheng. Geographical distribution and changes of peafowls in the historical period of China[A]. In: Wen Huanran, et al., Selected by Wen Rongsheng. Study on the changes of plants and animals in Chinese history[M]. Chongqing: Chongqing Press, 2006.

[44] Wen Xianji, Yang Xiaojun, Han Lianxian, et al. A survey on the distribution of green peafowl in China[J]. *Biodiversity*, 1995,3(1):46–51.

[45] Wen Yunyan, Xie Yichang, Li Xuehong. Study on the monitoring of green peacock in Konglonghe Prefecture Nature Reserve[J]. *Forest Survey and Planning*, 2016,41(4):69–71.

[46] Wu Jun. Investigation and analysis of the rearing and breeding status of green peafowl[J]. *Heilongjiang Animal Reproduction*, 2004,12(2):46–47.

[47] Xu Hui. Distribution status and protection measures of green peafowls in Chuxiong Prefecture[J]. *Yunnan Forestry Science and Technology*, 1995,72(3):48–52.

[48] Xu Hui. Status of the green peacock in Chuxiong, Yunnan[J]. *Chinese Journal of Wildlife*, 1999,20(3):12–13.

[49] Zheng Guanmei. Chinese pheasants[M]. Beijing: Higher Education Press, 2015.

[50] Zheng Zuoxin. Economic fauna of China • Birds[M]. Beijing: Science Press, 1963.

[51] Zhu Zhifei, Shen Manman, Cao Hui, et al. Observation and analysis on the behavior of blue peafowl during breeding period[J]. *Journal of Animal Science and Veterinary Medicine*, 2011(2):28–29.

[52] Zhu Shijie, Chang Hong, Zhang Guoping, et al. Analysis and phylogenetic analysis of mitochondrial cytochrome b gene of peafowl[J]. *Journal of Sun Yat-sen University*, 2004,43(6):45–47.

[53] Wang Shouchun. Study on Xinjiang peafowl name and origin of Kongquehe River[J]. *Western Region Research*, 2015(2):22–29.

[54] Wang Zijin. A Study on Qiuci "Peafowl" [J]. *Journal of Nankai Journal* (*Philosophy and Social Sciences Edition*), 2013(4):81–88.

[55] Wang Yanbo, Guo Fengping. The distribution of peacocks in China from the perspective of environmental history and its causes[J]. *Journal of Baoshan College*, 2018,37(1):29–36.

[56] Wang Yulong, Zhao Guangying, Tian Xiuhua, et al. Composition analysis and ultrastructure observation of egg shell of three peafowl species[J]. *Heilongjiang Animal Husbandry and Veterinary Medicine*, 2000,(4):36–37.

[57] Wang Zijiang. Yunnan peacock[J]. *Yunnan Forestry*, 1983(4):30.

[58] Wang Zijiang. The discovery of peacocks in Chuxiong, Yunnan[J]. *Zoological Research*, 1990,11(1):54.

[59] Wang Fang, Yao Chongxue, Liu Yu, et al. Investigation on the distribution of wild green peacocks in Xinping County based on infrared trigger camera technology[J]. *Forest Survey and Planning*, 2018,43(6):10–14111.

[60] Wang Fang, Jiang Guilian, Zhang Zhizhong, et al. Diversity and association analysis of associated birds and animals of green peacock in Xinping County, Yunnan Province[J]. *Chinese Journal of Wildlife*, 2018,39(4): 812–819.

[61] Xu Hui. Distribution status and protection measures of green peafowls in Chuxiong Prefecture[J]. *Yunnan Forestry Science and Technology*, 1995,72(3):48–52.

[62] Xian Fanghai, Yu Xiaogang. Studies on the biological characteristics and breeding habits of wild brodlehen in Tangjiahe Nature Reserve, Sichuan Province[J].*Sichuan Journal of Zoology*, 2008,27(6):1175–1178.

[63] Xie Yichang. Thinking of green peafowl protection in Konglonghe Nature Reserve[J]. *Forest Survey and Planning*, 2016, 41 (4) : 69–72.

[64] Xie Yichang. Current situation and protection countermeasures of biological resources in Konglonghe Nature Reserve[J]. *Forest Survey and Planning*, 2009,(1):10–12.

[65] Xinping Yi Dai Autonomous County Local Chorography Compilation Committee Office. Xinping Yearbook[M]. Dehong: Dehong Nationality Publishing House, 2017.

[66] Xu Long, Zhang Zhengwang, Ding Changqing. Application of line transect method in bird

population survey[J]. *Chinese Journal of Ecology*, 2003,22(5):127–130.

[67] Yang Lan. Ornithology of Yunnan, Volume 1, Non-passerines[M]. Kunming: Yunnan Science and Technology Press, 1995.

[68] Yang Xiaojun, Kong Dejun, Wu Fei, et al. The population status and protection of green peafowl in China[C]. The 13th National symposium on wildlife ecology and resources conservation and the 6th Western China Zoology Symposium, 2017–10–27

[69] Yang Xiaojun, Yang Lan. The observation of time budgets of captive green peafowl (*Pavo muticus*) [J]. Current Zoology, 1996,42(S1):106–111.

[70] Yang Xiaojun, Wen Xianji, Yang Lan. A survey on the distribution of green peafowls in the southeast and northwest of Yunnan Province [J]. *Zoological Research*, 1997, 18 (1):12, 18.

[71] Yang Xiaojun, Wen Xianji, Yang Lan, et al. A preliminary observation on habitat and behavior of green peafowl in spring[C]. Proceedings of the 4th Symposium on Ornithological Research in China, 2000,64–70.

[72] Yang Zhongxing, Wang Yong, Hua Chaolang, et al. Problems and countermeasures of green peafowl protection in Yunnan Province[J]. *Fujian Forestry Science and Technology*, 2019,46(4):112–119.

[73] John MacKinnon, Karen Phillipps, He Fenqi. A field guide of birds in China[M]. Changsha: Hunan Education Press, 2000.

[74] Yan Xiaojuan. Study on quality characteristics of blue peafowl meat[D]. Gansu Agricultural University, 2009.

[75] Yu Yuqun, Zhang Shanning, Gong Huisheng. Preliminary investigation on the distribution and density of pheasants in Foping Nature Reserve [J]. *Chinese Journal of Wildlife*, 1990,57(5):16–18.

[76] Yuan Jingxi, Zhang Changyou, Xie Wenhua, et al. A preliminary investigation of mammals and birds resources in Jiulianshan National Nature Reserve using infrared camera technology[J]. *Acta Theriologica Sinica*, 2016,36(3):367–372.

[77] Zhang Chunli. Feeding and reproduction of green peafowl[J]. Chinese Journal of Wildlife, 1995,(4):16–18.

[78] Zhang Lixia, Wei Zezhen, Wu Fei, et al. Artificial incubation and brooding of green peafowl[J]. *Feed Exposition*, 2015,(7):41–44.

[79] Zhang Yunmei, Wu Denghu, Yang Xiaoli, Pan Yongquan. Effects of ILT on blood chemical

indexes of peacocks[J]. *Journal of Animal Prevention and Control*, 2003,19(12):705–707.

[80] Zeng Zhaoxuan. On the extinction period of elephant, crocodile and peafowl in the Pearl River Delta Region[J]. Journal of South China Normal University (Natural Science Edition), 1980(1):173–185.

[81] Zhao Yuze, Wang Zhichen, Xu Jiliang, et al. Analysis of activity rhythm and time distribution of wild white-crened long-tailed pheasant by infrared photography[J]. Acta Ecologica Sinica, 2013,33(19):6021–6027.

[82] Zheng Guangmei. The Classification and distribution of world birds[M]. Beijing: Science Press, 2002.

[83] Zheng Guangmei. Classification and distribution of birds in China (3rd edition)[M]. Beijing: Science Press, 2017.

[84] Wang Qishan, Zheng Guangmei. Birds[A]. In: Wang Song, Ed. Chinese Red Book of Endangered Animals[M]. Beijing: Science Press, 1998.

[85] Zheng Zuoxin. A taxonomic list of Chinese bird species and subspecies[M]. Beijing: Science Press, 2000.

[86] Ministry of Forestry, PRC. National Key Protected Wildlife List[M]. Beijing: China Forestry Publishing House, 1989.

[87] Zhou Qingping, Chen Hong, Zhou Xuelin, et al. Determination of blood biochemical indexes of green peafowl[J]. Hubei Agricultural Sciences, 2011,50(15): 3124–3126, 3130.

[88] Zhou Xiaoyu, Wang Xiaoming, Jiang Zhenhua. Time distribution and activity rhythm of diurnal behavior of pheasants during overwintering in Helan Mountain[J]. Journal of Northeast Forestry University, 2008,36(5): 44–46.

[89] Zhu Ziqiang. Feeding management and reproduction of Peafowl[A]. In: Editor-in-Chief, Import and Export Office of Endangered Species, PRC. Domestication and Breeding of Endangered Economic Wildlife in China[M]. Harbin: Northeast Forestry University Press, 1997.

[90] Briekle N W. Habitat use, predicted distribution and conservation of green peafowl *Pavo muticus* in Dak Lak Province, Vietnam[J]. *Biological Conservation*, 2002,105:189–197.

[91] CITES. Convention on International Trade in Endangered Species of Wild Fauna and Flora[EB / OL]. https://www.cites.org/ （2020–02–01）.

[92] Dejun Kong, Fei Wu, Pengfei Shan, et al. Status and distribution changes of the endangered Green Peafowl (*Pavo muticus*) in China over the past three decades (1990s–2017) [J]. *Avian*

Research, 2018, 9(2):102–110.

[93] Dong F, Kuo HC, Chen GL, et al. Poputation genomic climat and anthropogenic evidence suggest the role of human forces in endengerment of green peafowl (*Povo muticus*)[J]. Proceedings of the Royal Society B, 2021, 288 (1948): 20210073.

[94] Fei Wu, De-Jun Kong, Peng-Fei Shan, et al. Ongoing green peafowl protection in China[J]. *Zoological Research*, 2019, 40(6): 580–582.

[95] Han L, Liu Y, Han B. The status and distribution of green peafowl *Pavo muticus* in Yunnan Province, China[J]. *International Journal of Galliformes Conservation*, 2007,(1):29–31.

[96] Howard R, Moore A. A complete checklist of the birds of the world[M]. London: MacMillan, 1984:109.

[97] Hurlbert S H. A coefficient of interspecific assciation[J]. *Ecology*, 1969, 50(1):1–9.

[98] IUCN. IUCN Red List of Threatened Species. [EB/OL]. http://www.iucnredlist.org. (2017–06–16)

[99] Jarwadi B H, Ani M, Hadi S A, et al. Behavior Ecology of the Javan Green Peafowl (*Pavo muticus* Linnaeus 1758) in Baluran and Alas Purwo National Park, East Java[J]. *HayatiI Journal of Biosciences*, 2011,18(4):164–176.

[100] Ke Y Y, Chang H, Zhang G P. Analysis of genetic diversity for wild and captive green peafowl populations by random amplified polymorphic DNA technique[J]. *Journal of Forestry Research*, 2004,15(3):203–206.

[101] Pudyatmoko S. Habitat and Spatio-Temporal. Interaction between green peafowl with cattle and megaherbivores in Baluran National Park[J]. *Journal of Forest Science*, 2019,13 (5):28–37.